Generation and Update of a Digital Twin in a Process Plant

Josip Stjepandić · Johannes Lützenberger ·
Philipp Kremer

Generation and Update of a Digital Twin in a Process Plant

Springer

Josip Stjepandić
PROSTEP AG
Darmstadt, Germany

Johannes Lützenberger
PROSTEP AG
Hamburg, Germany

Philipp Kremer
PROSTEP AG
Darmstadt, Germany

ISBN 978-3-031-47318-0 ISBN 978-3-031-47316-6 (eBook)
https://doi.org/10.1007/978-3-031-47316-6

This Springer imprint is published by the registered company Springer Nature Switzerland AG
The registered company address is: Gewerbestrasse 11, 6330 Cham, Switzerland

Paper in this product is recyclable.

The original version of the book has been revised by updating a few typographical errors and figure positions in the chapters. A correction to these chapters can be found at https://doi.org/10.1007/978-3-031-47316-6_11

Used Trademarks

CAMEO Enterprise Architecture	Dassault Systèmes
CAMEO Systems Modeler	Dassault Systèmes
AVEVA E3D Design	AVEVA Group
AVEVA Diagrams	AVEVA Group
CADMATIC	Cadmatic Oy
Hexagon	Hexagon AB

Contents

About the Authors

Josip Stjepandić (born 1961) looks back at an almost 40-year-long career in research and development of complex industrial products and services. Since 1996, he is working for PROSTEP AG, the leading consultancy for product data integration. He has conducted many projects on engineering collaboration, supplier integration, CAD data exchange, knowledge-based engineering, intellectual property protection, development of design methods, and systems engineering for leading global companies. Many of those works were published in the past years and presented at conferences.

Johannes Lützenberger (born 1984) graduated from the University of Bremen with Dipl.-Ing. in production engineering. During his over 10 years longing career, he gained various experiences with regard to Product Lifecycle Management, on technical as well as process levels, from academic and industrial perspectives. He was involved in and managed several projects dealing with semantical data integration, knowledge-based engineering, and software development. Since 2018, he is working at PROSTEP AG and currently holds the position of Product Manager for OpenDESC 3DigitalTwin. Outcomes of his work have been published and presented at conferences during the past years.

Philipp Kremer (born 1992) studied industrial engineering at the University of Siegen, with a special emphasis in mechanical engineering. During his studies, he gained insights into the processes and procedures of a renowned cable manufacturer. After completing his bachelor's degree, he transferred to TU Darmstadt for his master's degree. At the same time as his master's studies, he started working as a student trainee at PROSTEP AG, specifically in the area of systems engineering and the development of new technologies. After completing his master's degree, he is working permanently at PROSTEP AG.

Chapter 1
Introduction to the Digital Twin of a Process Plant

Abstract Recently popularized, as a term with no unique definition, digital twin refers to a high-fidelity digital replica of the physical object that updates and exchanges data synchronously with its physical counterpart. Data is continuously gathered and updated in real-time. A variety of digital twin types and expressions are used in research and practice for a wide range of use cases throughout the product lifecycle. The digital twin is used not only in technical domains. Its appearance is also registered in medicine, economy, and social sciences because the digital counterpart can be imagined for almost every artifact in the real world. The requirements for a digital twin are rigorous, especially if the digital twin of an existing system needs to be generated. In industrial practice, this means that the entire model must be manually rebuilt which induces costs and takes a long time. Considering its benefits, the digital twin is now obviously going to become an indispensable companion of a complex industrial system along its lifecycle which may last several decades. A process plant emerges from a laboratory level where the basic principles are developed, over the mini-plant level where the production parameters are adjusted, to the production plant level where the production occurs in the real environment. In plant engineering, the industry has been looking for an appropriate approach for a long time to generate the digital twin of a brownfield process plant. Moreover, process plants are subject to different, frequent changes during their long lifecycle. Examples are the implementation of new apparatus or the integration of new equipment in the process, as well as a change in the layout of the production. To avoid inaccuracies or inconsistencies during the retrofit, an efficient updating process is essential for an existing twin of a process plant. For this market segment, appropriate concepts and efficient solutions need to be elaborated and developed, including modern IT techniques. In this introductory chapter, the underlying approach for this book is drawn. The idea of the digital twin, the origins, the goals, the expected audience, the problem definition, and the objective of the work of this book are roughly described. Hence, we give the first insight into the content of this book and the mutual interdependence of the chapters. This book paves the way for how to generate and commercialize the digital twin of a process plant as a service in a collaboration scenario.

J. Stjepandić et al., *Generation and Update of a Digital Twin in a Process Plant*,
https://doi.org/10.1007/978-3-031-47316-6_1

Keywords Digital Twin · Process plant · Artificial intelligence · Continuous optimization · Cost reduction

1.1 Origins of the Digital Twin

Shortened cycles in product development and production, continuous optimization, and reduction of costs as well as the increase in quality of products and services must be combined and coordinated. In doing so, the challenge for companies is to make processes as cost-efficient as possible in order to achieve the greatest possible profit. The cost pressure and the competition between companies require a continuous improvement of overall efficiency [1].

In order to meet all these requirements, the digitization of processes is becoming increasingly important in modern industrial processes, in particular for those with a high share of manual tasks, such as production [2]. Digital twins [3] and artificial intelligence [4] have become mega-trends for this objective. The process industry has also recognized these trends and articulated its requirements [5].

The first occurrence of a digital twin is associated with Michael Grieves of the University of Michigan, who made the topic of digital twins the subject of his lecture on Product Lifecycle Management in 2003. Creating, managing, and using a virtual replica of a system (product, production system, or instance) throughout the whole system's lifetime (design, test, manufacture, and operation), is the background of the digital twin. Before the physical system is actually built, this holistic approach aims to complete a high-fidelity product definition, verification, and validation that includes the modes of failure. A digital twin results in the physical product being improved by preventing failures during use as well as better quality, better performance, lower costs, shorter time to market, and increased customer satisfaction [6].

In the context of this work, analysis, design, verification, and validation of the process chains take place. The modalities, dependencies, and conditions between the individual steps are elaborated. Based on the results obtained in this way, a concept for the generation of a digital twin of an existing process plant needs to be created as an extension of previous work [7]. In addition, problems that can occur in this process are identified. Considering its benefits, it becomes clear that the digital twin is going to become an indispensable companion of a complex industrial system along its lifecycle now [5].

The requirements for a digital twin are very high, especially when it comes to generating the digital twin of an existing system. In industrial practice, this means that the entire model must be manually rebuilt which induces costs and takes a long period. In plant engineering, the industry has been looking for an appropriate approach for a long time to generate the digital twin of a brownfield process plant [8]. Our approach, derived from a general manufacturing schema [2], responds to those requirements and is described here in a generic way.

In this chapter, the introduction to this book is presented: origins of the book (Sect. 1.2), goals of the book (Sect. 1.3), audience (Sect. 1.4), problem statement (Sect. 1.5), the objective of work (Sect. 1.6), and the content of the book (Sect. 1.7).

1.2 Origins of the Book

This book explores the results of the research and development projects on digital twins and artificial intelligence conducted in the past three years at PROSTEP AG, a globally acting software and consulting company headquartered in Darmstadt (Germany) [9]. This work was partially funded by public bodies. In the communication with the customers and partners, this work was recognized as particularly successful and worth further dissemination [10, 11]. Therefore, it was decided to present its outcome to a broader audience. As managers of the corresponding project teams, the authors express the recent developments in digital twins.

1.3 Goals of the Book

The focus of this book is an attempt to present the commercial applications and best practices of the digital twin of a process plant derived from both, the research work and the industrial projects led by the authors. This presentation comprises not only the most recent digital twin methods and techniques but also—and this is crucial—several real-world use cases from industrial companies and concepts for their commercialization. These examples and experiences are meant to demonstrate how the digital twin can facilitate process development and operations in the modern process industry.

The term "digital twin" refers to a wide range of concepts and methods that fall under the category of digital twin subtypes or digital twin derivates. Such archetypes are structured and assessed by using a compact taxonomy [12]. Each of these archetypes must be assigned to build a new product, process, or service. A substantial amount of collaboration and information exchange between individuals from other disciplines, functions, departments, or businesses must also be supported by appropriate methodologies and tools. In this book, we have discovered how this complex collaboration of globally distributed partners can happen in a productive environment.

First objective is to summarize achievements made possible using digital twins in the process industry and describe the current state of the art according to a comprehensive literature review. Secondly, discovering several options for structuring information flow in a modern process plant starting with requirements is a further objective. These options cover both organizational and technical methodologies and tool selection. A range of issues that must be resolved in practice and can be considered as a component of a customer-focused hierarchy is demonstrated by the approach and techniques. They ought to encourage making compromises and identify (almost) optimal solutions.

The third objective of this book is to demonstrate that digital twins have become essential, are widely distributed in many industries, and that new developing fields like autonomously optimizing plants can benefit from the same fundamental engineering concepts. The book's ultimate objective is to provide real-world examples and use cases that fully illustrate the possible achievements and benefits of a digital twin in real-world cases. This is accomplished through the presentation of concrete solutions for several plants of different purposes, sizes, constraints, and complexity. In addition, several current trends and practice-related issues have been identified.

1.4 Audience

The book's target readers include plant engineers, business planners, enterprise architects, postgraduate and PhD students, researchers, and software developers. It is also valuable for industry specialists, managers, and researchers. In this sense, the provided information is meant to be both a concise reference for more experienced specialists and an introduction to the creation and evaluation of novel digital twin approaches and methodologies.

Practitioners in these roles can utilize the information to brush their core competencies and refer to it as needed in the course of their daily activities, particularly while organizing commercialization actions. Graduate and undergraduate students who have mastered a number of fundamental engineering courses may find beneficial teaching material to put their skills to use in modern industrial design procedures. Researchers can find recent developments and practical challenges in several digital twin fields.

Managers need to grasp information reflecting different impacts and dimensions of digital twins for formulating a comprehensive strategy and implementing proper engineering structures and organization. They need to make decisions that enhance the company's competitiveness by meeting quality, cost, and timeline goals. This book opens the door for a workable approach to introducing digital twins. Management and engineering need to exchange information quickly and easily so that procedures may be modified to suit the business goals and so that the management can comprehend and monitor product issues and maturity. This book provides much information on organization, transmission, tracking, and tracing techniques.

The book fulfills the demand of students and researchers in the wide area of plant engineering who need comprehensive information on recent achievements and directions for future research. In this context, valuable information that can be used for this is provided in the last chapter of this book. Each chapter includes a comprehensive list of current, relevant sources that have either been used by the author or are suggested for further research.

Another potential audience consists of tool and platform developers, who typically have a good background in software engineering but a lower experience in the development of applications and processes. This book may be a helpful resource for people who create and implement integration scenarios, such as enterprise architects.

1.5 Problem Definition

Process plants are subject to different, frequent changes during their long lifecycle. Examples are the implementation of new apparatus or the integration of new equipment in the process, as well as changes in the layout of the production [13]. To avoid inaccuracies or inconsistencies during the retrofit, an efficient updating process is essential for an existing twin of a process plant [14].

Lifecycles of process plants last several decades. They emerge from a laboratory level where the basic principles are developed, over the mini-plant level where the production parameters are adjusted, to the production plant level where the production occurs in the real environment (Fig. 1.1) [15]. Most plants have a corresponding 3D CAD model. They provide the spatial information of a plant. However, they usually represent the planning status. Already during the construction of the plant, deviations occur. Since these are not documented or even digitized the models are no longer up to date.

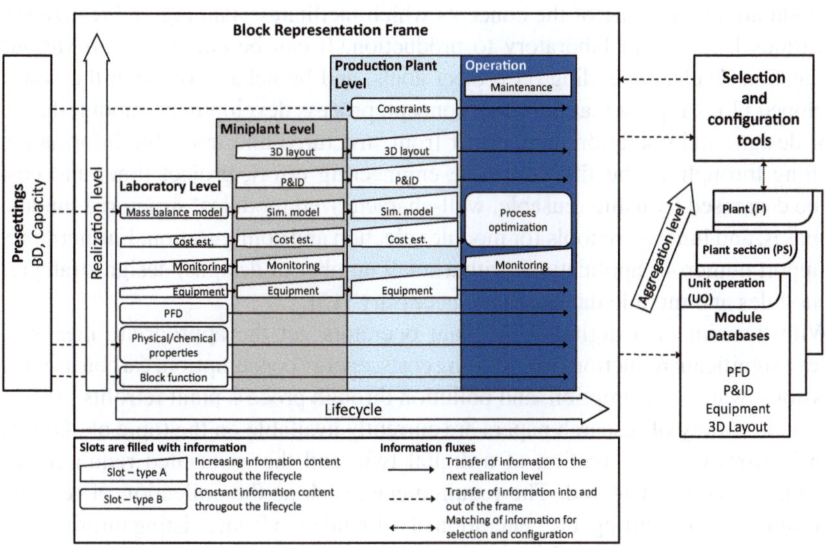

Fig. 1.1 Block representation frame of a process plant, derived from [15]

In contrast to the CAD models, the piping and instrumentation diagrams (P&IDs) of a plant are usually up to date, as this is required by law in most countries in the world. These P&IDs comprise information about the flow of the material as well as the linkage of the components on a logical level without any spatial information [16]. A 'block representation frame' stores' information and tracks information fluxes along the whole process and plant lifecycle and on different 'realization levels' from laboratory studies over mini-plant validation to production plant design and operation [15]. The different building blocks in Fig. 1.1 demonstrate the necessity for a tight coherence between process flow diagrams (PFD), P&IDs, and the 3D model of a plant.

The P&IDs represent the schematic, functional basis of a process plant. They are usually updated manually on hard copies. This way they are only human-readable and not digitized. Even though P&IDs are usually made available in PDF (not up-to-date), they lack in terms of machine-readability [16]. Digital twins can be created on this basis, but the costs of the manual effort are very high. Therefore, there is great interest in a digital twin that can be updated quickly and inexpensively by acquiring the physical twin in its current status [17].

Information handling should cover the specific views of all parties involved in the development process and production lifecycle (Fig. 1.1). Many important pre-settings and requirements for a development and investment project need to be considered. On the one hand, chemical product development, technical development of the process, and plant design are needed to be considered. On the other hand, aspects such as market analysis, business, and product strategy, as well as regulatory affairs come into place [15].

Modularization is one of the concepts which facilitates planning and realization on various levels from laboratory to production. It can be conducted on various aggregation levels, depending on project goals, and branches. By using the newly developed block representation frame concept, process development, module-based plant design, and operation can benefit from structured and traceable information handling throughout the lifecycle. The engineering effort, project time, and cost can be decreased by using reusable, well-characterized modules, structured module databases, and innovative tools for module selection and configuration. Furthermore, to support numerous applications with a small number of designs, design strategies for modules and module databases are necessary [15].

With the aim of a digital twin, plant operators get the appropriate means to achieve significant reductions in operating costs, energy consumption, carbon dioxide emissions, water consumption, and pollution through process plant retrofits [18].

A wide variety of research papers are currently available on the topic of "Digital Twins". However, the creation of a digital twin and the associated requirements have rarely been considered. The requirements, such as the collection of relevant information for retrofitting, the generation of a digital twin from existing information, and the minimum manual effort, are difficult to reconcile. In the sum of the entire factors, the creation of a digital twin turns out to be a complex undertaking with many risks [19].

1.6 Objective of Work

In this work, a concept is elaborated in which the orientation of digitization in process plants is presented. It aims at the consideration to accelerate the internal process and optimize the associated use of resources by means of a systematic approach. The creation of the concept presupposes basic principles, which will be explained in the following chapters in a generic order. Based on previous work, the underlying procedure was developed further (Fig. 1.2). It comprises three consecutive steps:

- Object acquisition by scans or similar techniques,
- Object recognition,
- Collection of models and data for digital twin.

The key point of this approach lies in the reliable recognition of relevant objects in a process plant based on scans. In chemical engineering process design, layout planning is a critical step in which safety factors must be considered. Layout planning considers both economic and safety constraints [20]. In previous studies, it has been stated that up to 80% of the total installation costs could be assigned to plant piping [21]. Therefore, the recognition of piping systems deserves particular attention.

Likewise, a way will be shown how the overall process can be automated to a high degree. In the process, methods based on a software application of artificial intelligence are used, which recognizes the relevant planning objects from a point cloud. In doing so, the artificial intelligence approach is based on an evaluation of the probability of an output, given an input [22]. After recognition, the objects can be transferred into CAD models. Optionally, a linkage with the corresponding P&IDs is envisaged.

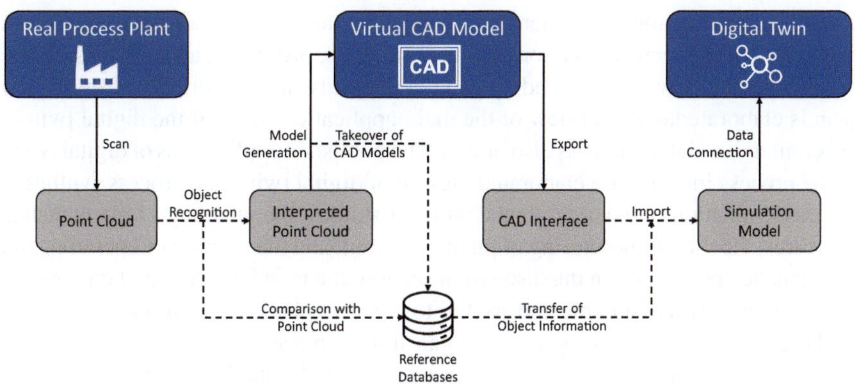

Fig. 1.2 General procedure to create a digital twin of a process plant, derived from [7]

1.7 Content of the Book

This book is subdivided into ten consecutive chapters, which will be introduced below. Singular chapters contribute to various aspects of basic concepts, methods, technologies, industrial applications, and current challenges of digital twins.

This chapter is this introduction to this book as explained in previous sections of this chapter.

Chapter 2 introduces the requirements and process design for digital twins of a process plant. The lifecycle of each successful system begins with a collection of its requirements which represent the expression of customers' needs. Requirements must be unambiguous, clear, unique, consistent, stand-alone, measurable, verifiable, and traceable. Functional and non-functional requirements are collected by using a systems modeling tool and subsequently enriched with further technical information. We extend the definition of requirements to the basic process which refers to the different phases of the product lifecycle: design, production, and operation. Therefore, this chapter formulates a consistent and detailed approach to define the requirements for digital twins of a process plant and derive a process model. In focus lies the transition of a requirements model, represented by a structure of functional and non-functional requirements in a matrix, to a flexible, multidimensional process in the Cameo Systems Modeler. Within the methodology explored in this chapter, a model to describe interdependencies between the tangible and intangible components as well as between the main contractor and subcontractor was implemented and the expected results were discussed including the different roles of two partners involved in the generation of the digital twin.

In the following Chap. 3, a comprehensive literature review of the digital twin of a process plant is presented. While it should be noted that a variety of definitions of the digital twin exist, several definitions of the digital twin are compared here. Countless publications document comprehensive approaches, advanced methods, and convincing practical solutions on how products, processes, and services can be optimized, and new ones created by using a digital twin. A taxonomy of the digital twin is elaborated. An overview of the main application fields of the digital twin in different industrial domains is also drawn. The published applications of digital twins in the process industry are elaborated: design of digital twins, the process synthesis, the modernization of existing plants, incident studies, the planning of maintenance measures, the virtual monitoring of plant operation, and the training of operating and maintenance personnel. In the discussion section, the match is presented on how the digital twin provides suitable approaches to business challenges a modern company in the process industry is facing. The gap in coherence between the real and the digital twin in a brownfield environment was detected. Finally, based on the digital twin characteristics, an approach for the generation and update of the digital twin of a process plant is drafted which should be offered as a commercial service.

Chapter 4 deals with the common practice in plant design with interoperability standards. Plant design as a phase of plant engineering comprises the activities in the conception and the planning of plants. Plant construction projects are usually defined as so-called engineering, procurement, and construction (EPC) projects which include the provision of the engineering design, the procurement of the necessary materials needed for construction, and finally the actual construction. EPC contractors are responsible for planning, designing, manufacturing/assembling, and handover the plant to the operator. In this context, the application scenario "Seamless and dynamic engineering of plants" (SDP) is explained, and the overall principle is presented as an initial engineering process for the engineering and construction of a plant. Ever deeper standardization in Industry 4.0 entails tool vendor-neutral representations of piping and instrumentation diagrams (P&ID) as well as 3D pipe routing. For the sake of the digital twin, a complete digital plant model requires combining these two representations. 3D pipe routing information is essential for building any accurate first-principles process simulation model, in particular for a brownfield plant. Piping and instrumentation diagrams are explored as the primary source for control loops. The interoperability as a fundamental capability of the digital twin is the key constituent of almost every initiative and approach.

In Chap. 5 relevant business cases for the use of digital twins of process plants are explained. The development of new offerings (products or services) belongs to a value-creation process. Such a process requires conceptual design and practical implementation. Hereby, the focus relies on how to create value for the customer. An accordingly defined business model is the key point. It describes the interdependencies between the product/service and its development. Especially for stakeholders in plant industries, that deal with complex assets, a digital twin providing 3D and spatial information respectively is of value. An automated generation of digital twins or at least a semi-automation reduces the costs to a reasonable level. According to the concept of an automated generation of digital twins with spatial information, current offerings in the market have been analyzed and reflected by applying the buyer utility map methodology. To shape a suitable business model according to the new offering, the approach of the business canvas has been exploited. The various fields to consider by developing the service solution, requirements from the target groups, the value proposition, and further have been defined that way. Finally, the customer's view of the concept of digital twin and use cases of the concept of digital twin including spatial information is presented and underlines the need for the mentioned service solution.

In Chap. 6, we discover the solution approach for the digital twin of a process plant and explore the conceptualization of such a solution. To find new business opportunities and involve new customer segments, increasing the market share, the servitization process is a fundamental means for manufacturing companies. In a working environment where only a few standard procedures exist, ideation—defined as the process of searching for a solution to a particular design issue—gets special importance. Based on a solution search, the product-service development is conducted from the initial requirements to the final customer service. The solution is drawn as a process chain with several steps which work like a 3D search of the components in the point cloud. A four-step methodology for the semi-automatic generation of digital twins from the point cloud of the plant is explored. By exploiting the Zachman framework the functional description of the legacy system is determined according to an already-used approach. Different actors (customer, capture partner, cooperation partner) in designed service during the emergence of a digital twin are described. The course of object recognition and the impact factors on the immediate and expected results are explained. Also, the importance of translation of the piping system description from the neutral to the proprietary environment (e.g., AVEVA) is emphasized.

Chapter 7 describes the implementation of a digital twin of a process plant. Adoption of different software components based on the previously presented conceptual solution and generate an integrating digital twin for a business workflow as a service extension. The focus of this software system lies in the implementation of methods from computer vision that can reliably recognize the existing objects in a process plant with their structure and interconnections. For several reasons, the implementation of an automatic object recognition procedure is realized by using existing convolutional neural networks (CNN), also known as deep learning. The recognition procedure is described in more detail, particularly how the piping system is built up in its full complexity coming from singular components. This recognition runs iteratively in four steps with a mutual interdependence. A robust clustering and structured aggregation yield a coherent pipe model before it is linked with piping and instrumentation diagram to form the digital twin. Due to the high complexity and variance of the tasks, some individual tasks must run fully manually or with partial human assistance. Finally, we present how the singular steps are automated and orchestrated by using a workflow automation platform.

In Chap. 8 the practical application of digital twins of a process plant is explored. Digitizing a brownfield process plant and the generation of a digital twin respectively is a challenging task. Object recognition from a point cloud highly depends on the quality of a point cloud. Environmental influences such as dust, dirt, vapors, and more are captured as well and need to be filtered out. Depending on the size and structure of the plants, several elements hurdle the scanning activities and lead to scanning shadows. In consequence, fragmentations appear, which reduce the quality of the results. Here, also the density regarding the packaging of the piping system and supporting constructive structures plays a role. Additional aspects such as insulation or painting change the appearance to some extent. While some components are simple connections between two components, others lead to a change of direction within the course of a piping system or even change attributes such as diameter. All that makes post-processing and optimization of parameters necessary. The practical application is presented in the context of examples from different domains. Various point clouds are analyzed according to the above-mentioned factors. Finally, the automatically generated piping systems are presented and approved in terms of integration with the original point cloud.

Chapter 9 deals with the creation of a new offering and explores the digital twin as a service. For continuous maintenance and repair activities, the as-is state of the specific brownfield plant needs to be digitized. Based on customer demands, new offerings for the generation of a digital twin of a process plant have been shaped. Input for the offerings are the point clouds as a result of laser scans of the specific asset. Depending on the size or complexity of these, the data volume increases drastically and decreases the manageability within CAD or authoring systems, respectively. The offer of Reduced Point Clouds addresses this use case. Within the *Basic Segmentation*, recognized objects are comprised in the categories pipeline, equipment as well as structure and are provided as single-point clouds. All together represent the entire scan, while the various categories/point clouds can be made visible or invisible. Further, the pipeline can be analyzed in more detail and categorized as pipe, elbow, tee, reducer, valve, or flange. Again, all are provided as single point clouds as well (*Segmentation Plus*). By clustering the points which describe one pipeline component, the structure can be broken down to the component level. Geometrical as well as spatial information can be derived. Based on that information, CAD models can be generated (*CAD Basic*). These models can be provided in the native format of specific CAD systems or in neutral formats like STEP or IFC with regards to BIM.

In Chap. 10, we present the conclusion and future trends in process plants. The generation of the digital twin for a brownfield process plant can be offered in the market as a service provided in collaboration with a few partners. It is a service with inherent self-improvement, while it facilitates a fast feed-forward of new software solutions, initiated by advances in research, to a daily-used industrial service. We draw the way for the further development of the 3DigitalTwin solution and highlight some still open challenges with respect to digital twins and their associated research (sub)domains in a brief overview. The precedent chapters, likewise, a consistent taxonomy with a self-explainable set of eight dimensions that can be articulated with a limited number of characteristics is reused for the assessment and positioning of current trends and challenges in digital twins. Four dimensions have been identified with potential for short-term advances: purpose and instantiation (reusability) of a digital twin, dealing with different accuracy of singular components, and improved processing of raw data. External impacts by advances in the scanning process, process modeling, and data processing and services are explored, too. A section is dedicated to the synchronization with the piping and instrumentation diagram which is identified as a key improvement for digital twin. In the discussion section, ten main challenges identified in the recent publications are addressed. Apart from the functional improvements, the handling of requirements, traceability, and metrics for the assessment of digitization pose the main challenges. Furthermore, the challenges in architecture, data processing, and automated simulation workflow round up this section. The importance of the protection of intellectual property in a collaborative environment is highlighted. An evolution of the digital twin's role from an enabler of cyber-physical systems to a constituent of Asset Lifecycle Management, the central platform for data integration and processing can be shown by the results. The conclusion section emphasizes again the importance of seamless interoperability.

References

1. Stjepandić J, Wognum N, Verhagen WJC (2015) Concurrent engineering in the 21st century. Springer International Publishing, Cham. https://doi.org/10.1007/978-3-319-13776-6
2. Stjepandić J, Sommer M, Denkena B (2022) DigiTwin: an approach for production process optimization in a built environment. Springer International Publishing, Switzerland. https://doi.org/10.1007/978-3-030-77539-1
3. Sharma A, Kosasih E, Zhang J, Brintrup A, Calinescu A (2022) Digital twins: state of the art theory and practice, challenges, and open research questions. J Ind Inf Integr 30:100383. https://doi.org/10.1016/j.jii.2022.100383
4. Goldman CV, Baltaxe M, Chakraborty D, Arinez J, Diaz CE (2023) Interpreting learning models in manufacturing processes: towards explainable AI methods to improve trust in classifier predictions. J Ind Inf Integr 33:100439. https://doi.org/10.1016/j.jii.2023.100439
5. Bamberg A, Urbas L, Bröcker S, Bortz M, Kockmann N (2021) The digital twin—your ingenious companion for process engineering and smart production. Chem Eng Technol 44:954–961. https://doi.org/10.1002/ceat.202000562
6. Grieves M, Vickers J (2017) Digital twin: mitigating unpredictable, undesirable emergent behavior in complex systems. In: Kahlen FJ, Flumerfelt S, Alves A (eds) Transdisciplinary perspectives on complex systems: new findings and approaches. Springer International, Cham, pp 85–113. https://doi.org/10.1007/978-3-319-38756-7_4
7. Sommer M, Stjepandić J, Stobrawa S, von Soden M (2023) Automated generation of digital twin for a built environment using scan and object detection as input for production planning. J Ind Inf Integr 33:100462. https://doi.org/10.1016/j.jii.2023.100462
8. Sierla S, Azangoo M, Rainio K, Papakonstantinou N, Fay A, Honkamaa P, Vyatkin V (2022) Roadmap to semi-automatic generation of digital twins for brownfield process plants. J Ind Inf Integr 27:100282. https://doi.org/10.1016/j.jii.2021.100282
9. NN (2023) PROSTEP AG. https://3digitaltwin.opendesc.com/. Accessed 20 April 2023
10. Grau M, Korol W, Lützenberger J, Stjepandić J (2021) Automated generation of a digital twin of a process plant by using 3D scan and artificial intelligence. Adv Transdisciplinary Eng 16:93–102. https://doi.org/10.3233/ATDE210087
11. Kremer P, Lützenberger J, Müller F, Stjepandić J (2022) An approach for the incremental update of a digital twin of a process plant. Adv Transdisciplinary Eng 28:310–319. https://doi.org/10.3233/ATDE220660
12. van der Valk H, Haße H, Möller F, Arbter M, Henning JL, Otto B (2020) A taxonomy of digital twins. In: Anderson B, Thatcher J, Meservy R (eds) Proceedings of the 26th Americas conference on information systems. pp 1–10
13. Sutton I (2015) Plant design and operations. Elsevier, Waltham, p 2015
14. Löwen U, El Sakka F, Schertl A, Fay A (2020) A systematic approach how to build, maintain and use an integrated plant model. Automatisierungstechnik 68(6):423–434. https://doi.org/10.1515/auto-2019-0097
15. Hohmann L, Kössl K, Kockmann N, Schembecker G, Bramsiepe C (2017) Modules in process industry—a life cycle definition. Chem Eng Process 111:115–126. https://doi.org/10.1016/j.cep.2016.09.017
16. Parisher RA, Rhea RA (2021) Pipe drafting and design, 4th edn. Elsevier, Amsterdam. https://doi.org/10.1016/C2019-0-01022-9
17. Davila Delgado JM, Oyedele L (2021) Digital twins for the built environment: learning from conceptual and process models in manufacturing. Adv Eng Inf 49:101332. https://doi.org/10.1016/j.aei.2021.101332
18. Uhlenbrock L, Jensch C, Tegtmeier M, Strube J (2020) Digital twin for extraction process design and operation. Processes 8(7):866. https://doi.org/10.3390/pr8070866
19. Semeraro C, Lezoche M, Panetto H, Dassisti M (2023) Data-driven invariant modelling patterns for digital twin design. J Ind Inf Integr 31:100424. https://doi.org/10.1016/j.jii.2022.100424

20. Anbari E, Parvini M (2022) Process plant layout optimization considering domino effects and safety devices in a large-scale chemical plant. Comput Chem Eng 167:108006. https://doi.org/10.1016/j.compchemeng.2022.108006
21. Latifi SE, Mohammadi E, Khakzad N (2017) Process plant layout optimization with uncertainty and considering risk. Comput Chem Eng 106:224–242. https://doi.org/10.1016/j.compchemeng.2017.05.022
22. Agapaki E, Brilakis I (2020) CLOI-NET: class segmentation of industrial facilities' point cloud datasets. Adv Eng Inf 45:101121. https://doi.org/10.1016/j.aei.2020.101121

Chapter 2
Requirements and Process Design for Digital Twin of a Process Plant

Abstract The lifecycle of each successful product, process, or service begins with a collection of its requirements which represent the expression of customers' needs. An appropriate requirements definition is therefore indispensable for high-quality results by preventing wrong turns and reducing the cost of change, both of prototypes and production tools, and ultimately the warranty costs. Requirements must be unambiguous, clear, unique, consistent, stand-alone, measurable, verifiable, and traceable. Coming from software engineering, a derived concept called application lifecycle management (ALM) has evolved rapidly as a new key constituent in the field of product lifecycle management (PLM). By definition of a tooled framework, the generation and tracing of a common metamodel related to a product, process, or service become possible. Functional and non-functional requirements are collected by using a systems modeling tool and subsequently enriched with further technical information. We extend the definition of requirements to the basic process which refers to the different phases of the product lifecycle: design, production, and operation. Therefore, this chapter formulates a consistent and detailed approach to define the requirements for digital twins of a process plant and derive a process model. In focus lies the transition of a requirements model, represented by a structure of functional and non-functional requirements in a matrix, to a flexible, multidimensional process model modeled in the Cameo Systems Modeler. Within the methodology explored in this chapter, a model to describe interdependencies between the tangible and intangible components as well as between the main contractor and subcontractor was implemented and the expected results were discussed including the different roles of two partners involved in the generation of the digital twin. Finally, future trends and needs for requirement engineering and its further integration with modern lifecycle management approaches are outlined.

Keywords Digital Twin · Requirements Engineering · Requirements modeling · Product Lifecycle Management · Application Lifecycle Management · Process model · Process modeling · CAMEO systems modeller

J. Stjepandić et al., *Generation and Update of a Digital Twin in a Process Plant*,
https://doi.org/10.1007/978-3-031-47316-6_2

2.1 Introduction

Requirements define in a formal manner the needs of stakeholders and what a desired good (e.g., product, process, solution) must provide to satisfy those needs. A requirement is [1]:

1. A condition or capability needed by a stakeholder to solve a problem or achieve an objective.
2. A condition or capability that must be met or possessed by a system or system component to satisfy a contract, standard, specification, or other formally imposed documents.
3. A documented representation of a condition or capability as in (1) or (2).

The formulation, documentation, and maintenance of requirements are the main objectives of requirements engineering (RE) [2]. It describes "*a process, in which the needs of one or many stakeholders and their environment are determined to find the solution for a specific problem*" [3]. Although the term RE is extensively used in software development, modern industrial development of products and services is indispensable without RE [4]. Missing or inappropriate RE causes misunderstanding among the stakeholders, budget overruns, insufficient functionalities, or even the cancelation or the final failure of the project [5]. However, the impact of RE is often ignored, yielding misunderstanding and errors in the specification and tracing of requirements. The assessment of requirements for criteria such as uniqueness, completeness, and verifiability remains neglected. Errors in requirements specification are regularly identified in later phases of development. The later they are discovered, the higher the costs of improving the errors [6]. That is why a need for a transparent and traceable way of following the requirements throughout the development process is indispensable.

Usually, RE is realized with the definition of a tooled framework that generates a common metamodel related to a product, process, or service [7]. This metamodel defines how the requirements and architecture of a system can be captured in such a way that they can be traceable from each other and origin specifications documents. If the interaction of process and product models is managed to address requirements identified in the preceding phases, then this model-based approach can support project management [8]. Coming from Software Engineering, Application Lifecycle Management (ALM) has evolved rapidly as a constituent of Product Lifecycle Management (PLM) which manages the requirements lifecycle as a non-material constituent of product structure [9].

High-quality product and process solutions presuppose a good requirements definition as a prerequisite which reduces the cost of change, both of prototypes and production tools, and ultimately the warranty costs. Thus, each development project starts with a requirements elicitation and negotiation phase in order to compile initial customers' needs into more formal expressed functional and non-functional stakeholders' requirements. Requirements are expressions of the customers' needs to identify and constrain a product or process. They must be unambiguous, clear, unique, consistent, stand-alone, measurable, verifiable, and traceable. Functional and non-functional requirements are enriched with technical information, transforming into technical, functional, or non-functional requirements [10].

Moreover, RE must consolidate requirements from every stage of the product or process lifecycle and feed the results back to the development process. Within systems engineering, RE propagates along the development process of a system and ensures a consistent and traceable elicitation and management of requirements. The ongoing interaction between RE and the development phases can be easily discovered in systems engineering by using the V-Model (Fig. 2.1) [2]:

Figure 2.1 illustrates the schedule and activities conducted during the individual phases of system development in separate layers. On the left side of the "V", stakeholder requirements are collected and decomposed into system requirements, from which the architectural design of the system is derived. After system implementation, the development results need to be tested against the initial specifications, which is done during the activities on the right side of the "V". Integration testing qualifies the correctness of the system architectural design, while the system test verifies the compliance of the whole system to the technical specification. Finally, an acceptance test validates that the system meets the needs of the stakeholders [2].

This chapter aims at giving an overview of how to collect requirements set in the context of the digital twin and how to subsequently derive the process model to reliably generate a digital twin of a process plant. Section 2.2 describes how the RE approach is applied to extract, verify, weigh, and prioritize the singular requirements and how a set is created. Section 2.3 presents a generic approach to how to model the process based on a set of requirements. Section 2.4 finally summarizes the results and gives perspectives with regard to the identified gaps and challenges.

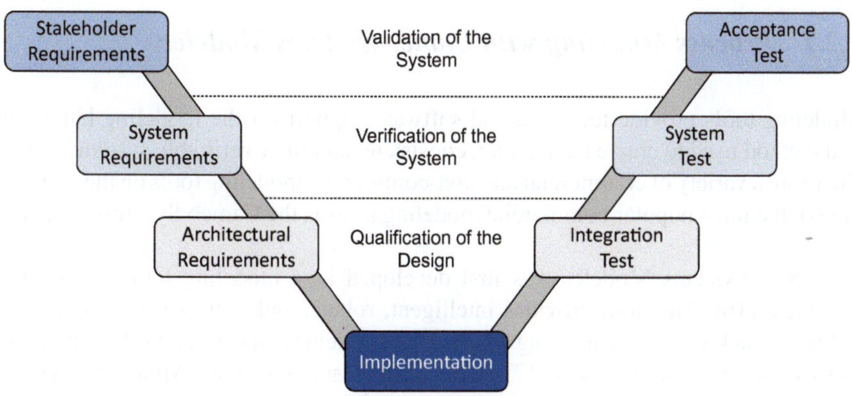

Fig. 2.1 Requirements engineering in the V-Model, according to Hull et al. [11]

2.2 Requirements for the Concept Twin

Requirements can be related to capabilities needed by certain stakeholders, or capabilities that must be provided by a system. In this sense, requirements in the system development process can be assigned to two distinguished areas: the problem domain and the solution domain [12]. The problem domain includes the needs and business goals for system development and their formulation into stakeholder requirements, without preselecting any specific solution characteristics. The solution domain contains the system requirements describing the targeted functionalities of the solution and subsequently the architectural design, which specifies how the solution will meet the system requirements [2]. In this work, we will focus on the system requirements in the solution domain which begins with requirements elicitation.

The stakeholders of the overall process need to be identified now and the individual elements explained in more detail. In this section, the requirements for the digital twin that is to be created at the end of this concept are elaborated based on previous knowledge and interviews with the experts. For this purpose, a list of requirements is first drawn up and these requirements are divided into desirable and mandatory requirements. After the corresponding subdivision, the requirements are formulated. From these, a Derived Requirement Matrix [13] is provided afterward with the software tool Cameo Systems Modeler [14]. Thus, it is to be visualized, how requirements are connected in a hierarchy. Subsequently, the matrix is evaluated, and the results are presented. To give the reader a better understanding, excerpts from the respective lists are shown in this section. The creation of the requirements lists is carried out according to Anderl et al. [15].

2.2.1 Process Modeling with Cameo Systems Modeler

Modeling tools provide technical and software support for the modeling language and method used to create a complete, consistent, and fully verifiable system model. There are a variety of commercial and non-commercial modeling tools on the market. One of the most popular commercial modeling tools is the Cameo Systems Modeler [14].

Cameo Systems Modeler was first developed as a modeling tool in 2007 by NoMagic [16]. This tool provides intelligent, robust, and intuitive procedures for defining, tracking, and visualizing all aspects of systems in most standards-compliant SysML models and diagrams [17]. Meanwhile, Cameo Systems Modeler emerged as a cross-platform collaborative multi-purpose Model-Based Systems Engineering (MBSE) environment.

Consistently performing model-based system modeling with an appropriate temporal view, structural view, and system architecture, including linking the functional and logical elements, provides a system model with all relations of the system elements. Accordingly, the system model can be used by means of SysML in Cameo

Systems Modeler to generate a complete functional structure and a complete system architecture.

In addition, the activity diagrams created allow the causal relationships of the functions to be represented, so that a systematic cause-and-effect analysis can be performed with Cameo. This enables a systematic identification of all fault conditions and their causes. In addition, the physical and system dynamic properties can be modeled and simulated. The results provide an indication of which concepts can best meet the previously defined requirements. Optimization potentials are also identified in the early phases. This makes Cameo an optimal tool to support traceability, validation, and decision-making for the development of complex systems [18].

To support the simulation of behavioral diagrams, Cameo uses Simulation Toolkit. The toolkit is based on the standards of OMG fUML (Foundation Subset for Executable Models) and W3C SCXML (State Chart XML). This is what makes simulation possible in the first place. In addition, a specific Graphical User Interface (GUI) can be created with Cameo Systems Modeler. Its simulation is based on Java Swing. When starting such a simulation a corresponding Java Swing User Interface is generated. This suggests that also for the simulation of the SysML behavioral diagrams executable Java objects are generated at runtime and are involved in the animation [16].

For our purpose, Cameo Systems Modeler will be used (a) to run engineering analysis for the design decisions evaluation and the requirements verification, (b) to continuously check the model consistency, and partially (c) to track the design progress with metrics.

2.2.2 Requirements Elicitation and Validation

As a starting point, a bulk list of requirements for the digital twin was created based on the previous experience in OpenDESC.com, the interviews with the prospective customers, and the literature recherche [19]. Then a division was made into desired and mandatory requirements. Overall, the list is divided into 13 categories of requirements, with a total of 101 requirements elaborated. The requirements relate to the technical and business areas of the digital twin. The categories are briefly explained below:

1. **Recognition of the piping components from a point cloud**: this category includes the requirements that the individual objects are to be recognized from a point cloud [20]. The objects to be recognized will be presented and explained in more detail in Chap. 6.
2. **Clear position and orientation of objects in space**: objects can be described by different possibilities in space. This category of requirements includes information about the center lines as well as about the center point of the bounding box of each component [21].

3. **Clear determination of geometries of objects**: This category includes information about possible geometry properties of the objects (such as the diameter, length, height, or width of an object) [21].
4. **Correct naming of components**: This category groups the requirements that correctly and uniquely name a component. For example, it is the individual designation of each component [21].
5. **Assignment of attributes to components**: The category refers specifically to the piping components. Thus, information about the insulation as well as the colors of the pipe classes must be correctly assigned [21].
6. **Enrichment of the process values**: There are requirements for the digital twin that the information from the measurement systems must be read out correctly and then processed. The requirements relate in particular to the pressure, temperature, and flow measurements [21].
7. **Generation of CAD models**: If an object already exists in a model library, it should also be correctly extracted. These requirements refer primarily to the equipment, of which a CAD model already exists [21].
8. **Correct classification of the error information**: Error information must be incorporated into the digital twin after correct classification without errors. In this way, the digital twin can continue to develop and improve [21].
9. **Product information**: The digital twin should contain information about the product. These are the characteristics, the application possibilities, as well as the technical data of the products.
10. **Correct identification and filtering of the disturbance factors**: The disturbance factors that influence a digital twin can be identified and filtered. These are requirements that relate to the environment and surroundings as well as the structuring and safety of a process plant.
11. **Maintenance and service of the infrastructure**: The category includes the requirements that the digital twin can be operated with as little effort as possible for the hardware and software [22].
12. **Handling of the Digital Twin**: This category of requirements relates to the handling of the digital twin. On the one hand, handling should be as simple as possible for authorized individuals. On the other hand, the digital twin must be protected from access by unauthorized people [23].
13. **Economic requirements**: The requirements in this category relate to the economic factors of a digital twin [24]. The requirements for the total costs and the profit to be generated by the digital twin project are explicitly formulated.

At this point, a brief overview of the categories into which the requirements are divided takes place. The following is an excerpt from the list of requirements so that the reader gets an overview of how the Excel file is structured (Fig. 2.2).

The header describes the most important information about the requirements list. The left part of the header contains information about the company and the department for which the requirements list was created. In addition, the name of the person who created the requirements list can be found there. In the middle of the header, there is the designation or name of the project. On the right side of the header is the

Company:	PROSTEP AG				**List of Requirements for the Digital Twin of a Process Plant**	Creation Date: 07.02.2022
Department:	Opendesc.com					ID: DigiTwin_Plant_Engineering
Editor:	Philipp Kremer					
ID-number:					Requirement:	Desired / Fixed
1					**Clear Identification of the Objects**	
	1				Piping parts	
		1			Pipe	D
		2			Fittings	
			1		Elbow	F
			2		Reducer	F
			3		Tee	F
			4		Intersection	F
			5		Plug	D
		3			Connection elements	
			1		Flange	F
			2		Sealing	
		4			Armatures	
			1		Valves	
				1	manual	D
				2	automatic	D
		5			Pipe support	D
	2				Equipment	
		1			Machines	
			1		Working machines	
				1	Pumps	D
			2		Power machines	

Fig. 2.2 Section of the list of requirements for the digital twin

classification and identification information. As shown in the figure, the categories into which the requirements are classified start with "1" and have a yellow background. The requirements in the categories are then assigned to the corresponding ID numbers. The right column describes whether the requirement is desired or fixed. Next, the requirements are formulated and described with the respective ID numbers. The next figure shows an excerpt of the requirements written out from the EXCEL file.

In Fig. 2.3, the first column from the left contains the name of the requirement. Next to the names of the requirements, a column with the corresponding ID numbers is inserted. The wording of the desired and mandatory requirements is in the "Text" column. The desired requirements can be recognized by the fact that a "should" wording has been chosen. The fixed requirements, on the other hand, were formulated with a "must". In the last column, the ID numbers are inserted, from which the requirements can be derived. This is the basis for creating the *Derived Requirement Matrix*. The matrix is created with the software tool *Cameo Systems Modeler*. In this way, the formal description of requirements can also be used for the simulation of the impact of desired requirements. For this purpose, the list with the formulated requirements is imported into the program and the matrix is created. A section of the matrix is shown by the term in figure 2.3.

Figure 2.4 shows a section of the Cameo Systems Modeler application. The requirements are listed once from top to bottom and from left to right according to the ID numbers. The arrows in the cells represent a corresponding derivation. The solid arrows symbolize a direct derivation from the requirement, while the dashed arrows symbolize an indirect derivation. Directly next to the requirements, the sum is shown in how many requirements are derived from the respective requirement.

Name	ID-Number	Text	Derived Requirment
Clear identification of the objects	1.		
Piping Parts	1.1.		1.
Pipe	1.1.1.	The pipes should be identified clearly from the point cloud.	1.1.
Fittings	1.1.2.		1.1.
Elbow	1.1.2.1.	The elbows must be identified clearly from the point cloud.	1.1.2.
Reducer	1.1.2.2.	The reducers must be identified clearly from the point cloud.	1.1.2.
Tee	1.1.2.3.	The tees must be identified clearly from the point cloud.	1.1.2.
Intersection	1.1.2.4.	The intersections must be identified clearly from the point cloud.	1.1.2.
Plug	1.1.2.5.	The plugs should be identified clearly from the point cloud.	1.1.2.
Connection elements	1.1.3.		1.1.
Flange	1.1.3.1.	The flanges must be identified clearly from the point cloud.	1.1.3.
Sealing	1.1.3.2.	The seals must be added automatically to the Digital Twin.	1.1.3.
Armatures	1.1.4.		1.1.
Valves	1.1.4.1.		1.1.4.
Valve manual	1.1.4.1.1.	The manual valves should be identified clearly from the point cloud.	1.1.4.1.
Valve automatic	1.1.4.1.2.	The automatic valves should be identified clearly from the point cloud.	1.1.4.1.
Pipe support	1.1.5.	The pipe supports should be identified clearly from the point cloud.	1.1.
Equipments	1.2.		1.
Machines	1.2.1.		1.2.
Working machines	1.2.1.1.		1.2.1.
Pumps	1.2.1.1.1.	The pumps must be identified clearly from the point cloud.	1.2.1.1.

Fig. 2.3 Section of the table with formulated requirements

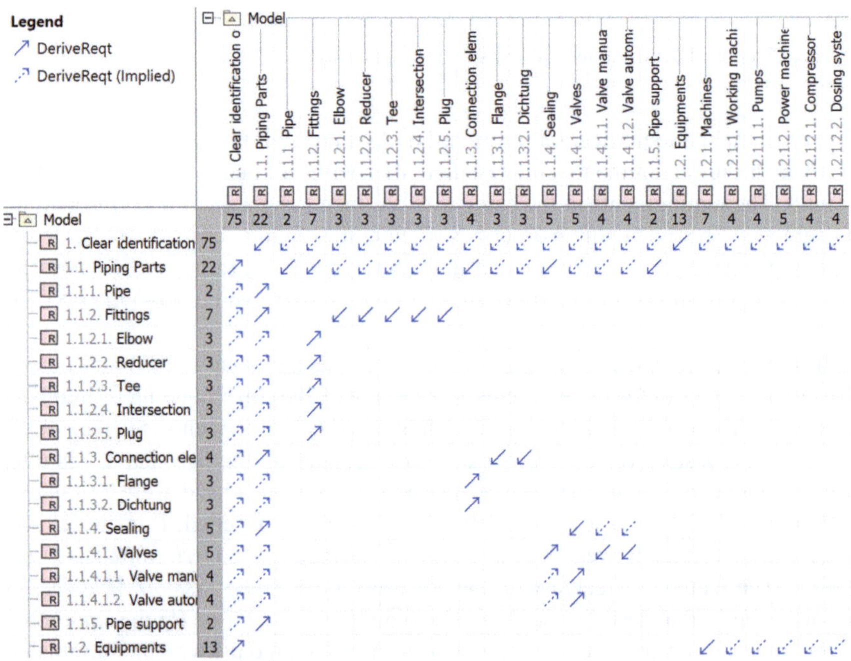

Fig. 2.4 Section of the derived requirement matrix

To check and evaluate this matrix within the requirements evaluation, it is exported as a CSV file which makes further processing easier, analogous to the dependency structure matrix (DSM). DSM provides a simple and concise way to represent the deconstruction of a complex system. It is amenable to powerful analyses, such as clustering (to facilitate modularity) and sequencing (to minimize cost and schedule risk in processes) [25]. These capabilities will be used here.

The matrix is reformatted in EXCEL so that each arrow is marked by a "1" (Fig. 2.5). This provides a better overview and gives the input for matrix calculation methods. In addition, the counting and the prioritizing of the derived requirements are simplified by standard functions. In Fig. 2.5, a section of the file is shown, which should help to evaluate the matrix.

Derived Requirement / Requirement	Percentage of derived requirements	Number of derived requirements	**1. Clear Iden**	1.1. Piping Pa
1. Clear Identification of the Objects	**71**	**75**		1
1.1. Piping parts	21	22	1	
1.1.1. Pipe	2	2	1	1
1.1.2. Fittings	7	7	1	1
1.1.2.1. Elbow	3	3	1	1
1.1.2.2. Reducer	3	3	1	1
1.1.2.3. Tee	3	3	1	1
1.1.2.4. Intersection	3	3	1	1
1.1.2.5. Plug	3	3	1	1
1.1.3. Connection elements	4	4	1	1
1.1.3.1. Flange	3	3	1	1
1.1.3.2. Sealing	3	3	1	1
1.1.4. Armatures	5	5	1	1
1.1.4.1. Valve	5	5	1	1
1.1.4.1.1. Valve manual	4	4	1	1
1.1.4.1.2. Valve automatic	4	4	1	1
1.1.5. Pipe support	2	2	1	1

Fig. 2.5 Excerpt of the evaluation file

The requirements are also listed in the evaluation file from top to bottom as well as from left to right. The matrix can be read from left to right. Starting from the requirements in the first column, the requirements that are derived from this requirement are marked with a "1" on the right. The requirements that are direct and indirect derivation are considered. In the "Number of derived requirements" column, the ones are counted out. The percentage of derived requirements is calculated according to the following formula:

$$\frac{Number\ of\ derived\ requirements}{Total\ number\ of\ requirements} * 100$$

By calculating the percentage of derived requirements, it is now possible to analyze from which requirement the most requirements are derived. This allows the evaluation, in particular in case of changes. The following Table 2.1 shows the requirement categories with the corresponding percentage shares of the derived requirements.

Table 2.1 Requirements with percentages

Number	Requirement categories	Derived requirements (%)
1	Clear identification of the objects	71
2	Clear position and orientation of objects in space	13
3	Clear determination of geometries of the objects	5
4	Correct naming of the objects	4
5	Correct assignment of attributes of the objects	7
6	Correct processing of information from the measurement systems	5
7	Correct extraction of objects from a model library	3
8	Correct classification of the error information	3
9	Product information	4
10	Correct identification and filtering of disruptive factors	3
11	Infrastructure maintenance	2
12	Handling of the digital twin	3
13	Financial requirements	4

2.2.3 *Results of Requirements Validation*

Requirements validation aims to provide proof that the requirements specification adequately meets the needs of the stakeholders. It is used to confirm the completeness and correctness of the determined requirements, in order to ensure that the documented requirements accurately express the stakeholder's needs [2]. Therefore, stakeholders must participate in requirements reviews during the validation process to make sure requirements are suitable for the product and approve the final result.

As presented in Table 2.1, exactly 71% of the total requirements are derived from the requirement category "Identification of the objects" and are homogenous in all stakeholder groups. Between 2% and 13% of the requirements are derived from the remaining categories. A concept can now be created based on the analysis of the requirements. For this purpose, the concentration is placed on the requirements from which the most requirements are derived (identification, localization, and determination of geometries of the object). As will be presented later, the importance and the weight of different requirements related to the singular object categories remain disparate. Nevertheless, the requirements must consider all stated business objectives and must be expressed clearly and understandably so that any missing requirements can be identified.

2.3 Process Design

Following the modern model-based systems engineering techniques, the processes are designed based on the identified stakeholders and the requirements for the concept twin. To this end, the cooperation process and the internal process are presented and explained in more detail from the perspective of a general contractor. Consistent data modeling enables simple changes across all affected tools and easy verification of preliminary results. Redundant information can be identified and eliminated, and networking between different data is enabled. Inconsistency in component management can be analyzed and addressed at different hierarchical levels (pipe, pipe system, entire plant) [26]. The processes are then illustrated and verified in *Cameo System Modeler* using a pseudo code [27]. Of course, this facilitates the tracing of requirements which can be achieved by introducing support for capturing, structuring, and mapping between decisions and resulting outputs, such as piping building blocks, knowledge implemented as rules, and the argumentation for the selection, design, and specification of these [28].

2.3.1 Cooperation Process

Complex development projects such as the generation of a digital twin can involve dozens of engineers. It is a common practice to outsource such activities to service providers. In our case, it was planned that the general contractor provides the main process with the support of a partner who contributes to the specific tasks related to plant engineering. In the previous sections, the basis for the concept was created. An overview of the elements in the overall system was provided. In this section, the process between the main contractor (PROSTEP AG) and a cooperating partner is considered in more detail. This cooperation process is shown as an activity diagram in the following figure. The activities symbolize the individual process steps and are subsequently explained in more detail.

Figure 2.6 shows the cooperation process between PROSTEP AG and a cooperating partner as an example. The following is an explanation of the process steps. First, the steps on the PROSTEP AG side are explained up to the "Generate XML" activity. It should be noted that only activities from the diagram are described in which PROSTEP AG is also actively involved. The activities on the part of the cooperating partner will not be described in any further detail:

- **Extraction Point Cloud from the Exchange Memory**: The scanned point cloud of the plant is transmitted to PROSTEP AG via an exchange medium. This point cloud is the input for the internal process and must be prepared. The point cloud is transmitted in the *.e57 format.
- **Segmentation**: The input for the activity is the processed point cloud from the previous step. A distinction is made in the internal process between the 1st and 2nd segmentation, whereby further segmentation levels are quite conceivable for

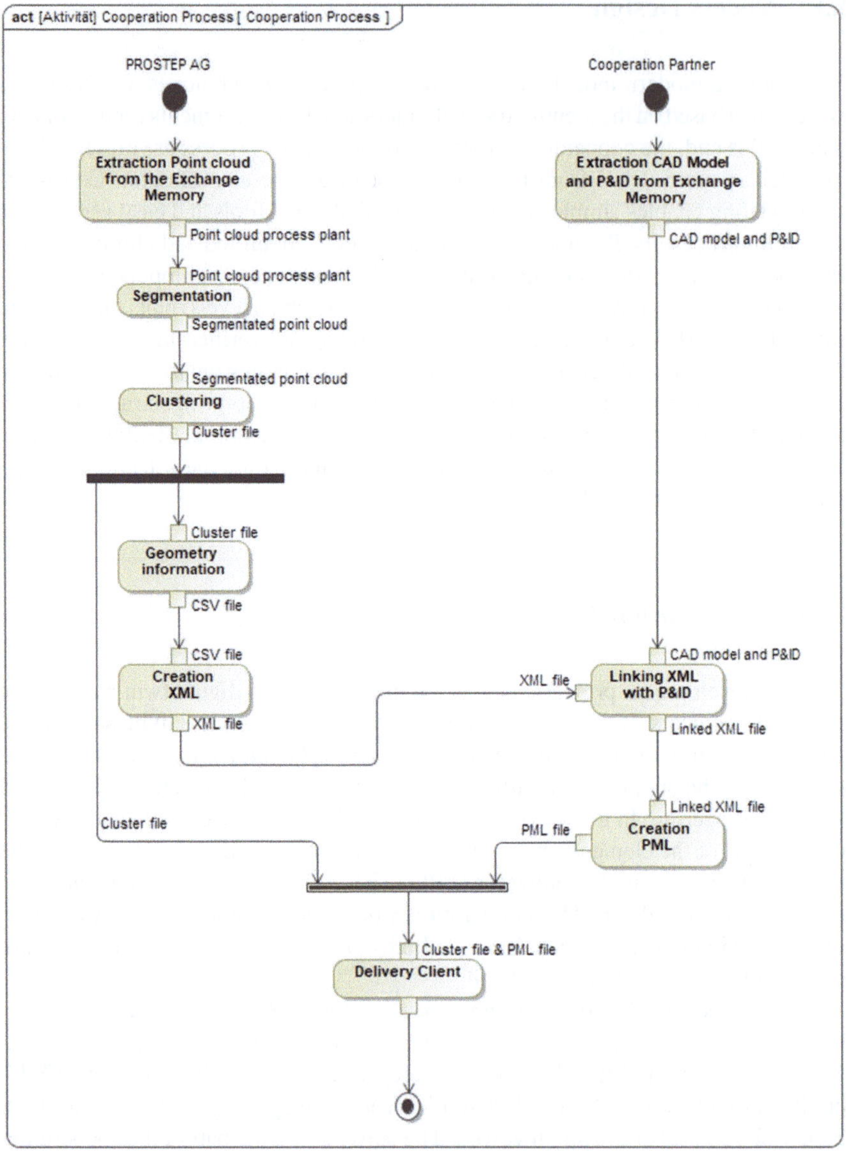

Fig. 2.6 Cooperation process

complex systems. For this, reference is made to the section "Internal process", where the segmentation step is explained in more detail. In general, it can be said that in this step the processed point cloud is divided into individual segments which comprise specific classes of components and parts. The output of this segmentation is the point cloud logically divided into several segments.

- **Clustering**: This segmented point cloud is the input for the next process step, clustering. Here, all points are assigned to a group or an object that has a distance from each other below a certain value. This value should not be below the tolerance from the point cloud, otherwise, the points cannot be properly assigned to an object. Compared to mechanical engineering, smaller tolerances are assumed in plant engineering. Thus, the selection of the tolerance depends very much on the industrial sector in which the digital twin is to be built. The output of the clustering is the clustered point cloud in a cluster file.
- **Geometry information**: Based on the cluster file, the geometric information of the components (e.g., diameter/radius) and position information (e.g., center lines and center of the bounding boxes of each component) can be derived. For the sake of easy handling, this information is derived as a CSV file.
- **Creation XML**: With the CSV file and the corresponding geometry information, an XML file is generated in this process step. For this purpose, PROSTEP AG and its cooperating partner agreed on which information is required for the internal process at the respective companies.
- **Extraction CAD model and P&ID from the Exchange Memory**: In this process step, the cooperating partner takes the CAD model and the P&ID from the exchange vault, which was provided by the customer.
- **Linking XML with P&ID**: The purpose of this step is to link the information contained in the P&ID with the detected 3D objects from the point cloud. For this purpose, the topology of the P&ID is matched with that of the scan and a link is created. In this way, the recognized objects can be enriched with the information from the P&ID. Input for this process step are the clustered point cloud, the CAD model, and the P&ID.
- **Creation PML**: Piping Modelling Language (PML) is the native piping format in the AVEVA system world. As already described, the focus is initially on the feedback of the models into AVEVA. Based on the position coordinates from the point cloud as well as the meta information from the P&ID, a PML file is created. This can be read by the customer into e.g., AVEVA E3D system and visualized automatically. Other system providers are also to be served in the future but are not considered further in the context of the initial project.
- **Delivery Client**: The clustered point cloud as well as the generated PML file are transmitted to the customer. On this basis, a plant redesign is possible.

2.3.2 Internal Process PROSTEP AG

The following Fig. 2.7 shows an activity diagram of the internal process at PROSTEP AG. Here, too, the activities are representative of the individual process steps. These are explained in more detail below.

The internal process consists of the steps described below. Individual steps overlap with the cooperation process and could be repeated if required by the cooperation

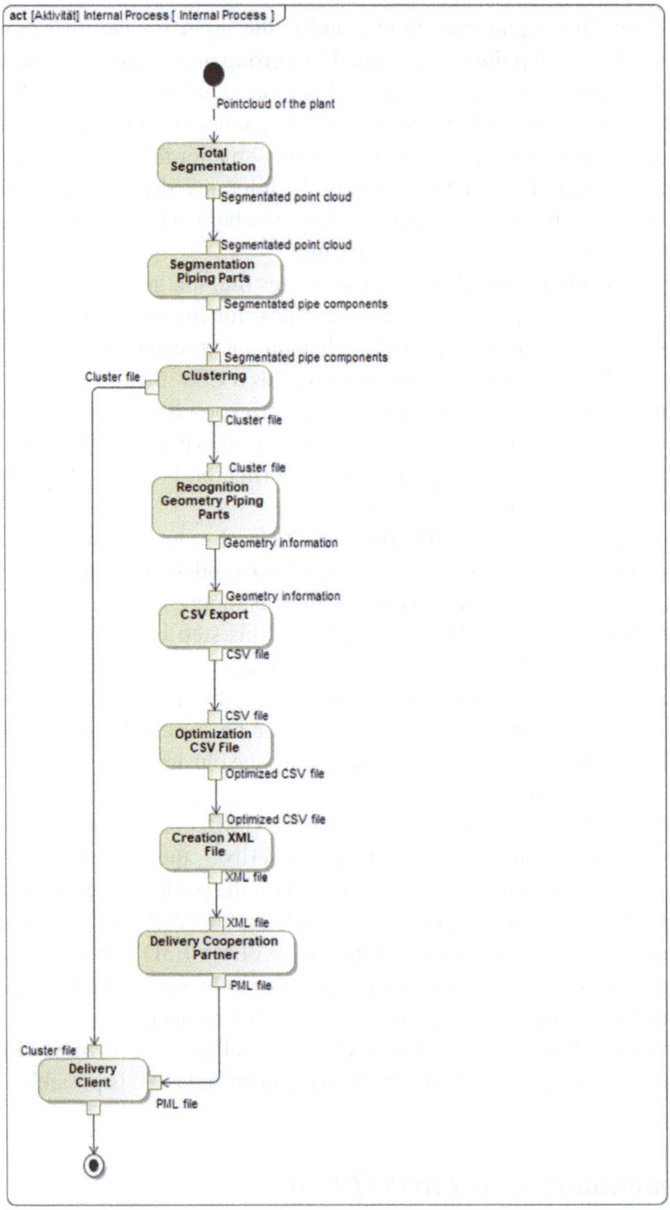

Fig. 2.7 Internal process

process. The point cloud of the process plant is used as the input of the internal process.

- **Total Segmentation**: The point cloud of the process plant is divided into individual segments in this process step. The output of this process step is the subdivision of the point cloud into the segments pipeline, equipment, and structure. Further segments (e.g., auxiliary equipment) could be provided if required by the customer.
- **Segmentation Piping Parts**: The segmented point cloud of the "pipeline" serves as the basis for this process step. In this, the point cloud "pipeline" is subdivided into different segments or different piping parts. At the end of this process, the components of a pipeline system (pipe, elbow, flange, valve, throttle, pump) can be viewed differently from each other.
- **Recognition Geometry Piping Parts**: In this process step, the geometries of the previously segmented pipeline components are recognized. This includes information about the center lines and the center of the bounding boxes of the respective components. In parallel, the components of the equipment are also recognized according to similar rules.
- **Clustering**: In this process step, the points of the segmented cloud are assigned to the individual clusters. The output of this activity is the individual component clusters with the corresponding geometric information. The different clusters are displayed in different colors to view the components in a more differentiated way.
- **Delivery Client**: The component clusters with the corresponding geometric information are also called cluster files. This file is provided to the customer for further processing. With this file, the customer has an overview of which parts occur and how often in the process. This is important in order to compare whether the built plant actually matches the planned plant. With the handover of the file to the customer, the first output of the internal process is created.
- **CSV Export**: Based on the cluster file, the geometry information is transferred to a CSV file. In this file, the information of the individual components is summarized. This includes e.g., information about the center lines as well as the information about the center of the bounding box.
- **Optimization Geometry File**: In this process step, the generated geometry file is optimized. For this purpose, the positions of the individual components are corrected if there are deviations.
- **Delivery cooperating partner**: The geometry file created is transmitted with the meta information to the cooperating partner in this process step.

2.3.3 Verification of Concepts

In the previous section, the cooperation process and the internal process at PROSTEP AG were elaborated, presented, and explained. These processes are illustrated and verified in this section using pseudocodes (Fig. 2.8). In this way, it can be shown that the concept meets with its approval. The pseudo codes are written with the application program “Notepad++”. In the following figures, the pseudocodes are illustrated and then explained.

In line one of the verification procedure for the cooperation process, the first PROSTEP class is opened. Subsequently, the process steps from the cooperation process that are executed by PROSTEP AG are defined as functions among each other: the extraction of the point cloud from the exchange memory, the segmentation, the clustering, the geometry information, the generation of XML, and the delivery to the client.

```
1   begin class PROSTEP
2
3   function ExtractionPointCloudFromtheExchangeMemory (Input PointCloudProcessPlant){
4       //Point cloud is extracted from the exchange memory
5       return PointCloudProcessPlant
6   }
7   function Segmentation(Input PointCloudProcessPlant){
8       //Segmentation point cloud
9       return SegmentatedPointCloud
10  }
11  function Clustering(Input SegmentatedPointCloud){
12      //Clustering of the segmentated point cloud
13      return ClusterFile
14  }
15  function GeometryInformation(Input ClusterFile){
16      //Derivation of the geometry information (e.g. diameter)
17      return CSV-File
18  }
19  function CreationXML (Input CSV-File){
20      //Generation of the XML from the CSV file
21      return XML-File
22  }
23  function DeliveryClient(Input ClusterFile){
24      //Cluster file is delivered to the client
25  }
26
27  end class
28
29  begin class CooperationPartner
30
31  function ExtractionCAD-ModelAndPIDfromExchangeMemory (Input CAD-Model, PID){
32      //Extraction of the CAD model and the P&ID from the exchange memory
33
34  function LinkingXMlwithPID(Input XML-File, CAD-Model, PID){
35      //Linking the XML file with the CAD model and the P&ID
36      return LinkedXML-File
37  }
38  function CreationPML(Input LinkedXML-File){
39      //PML file is derived from the linked XML file
40      return PML-File
41  }
42  function DeliveryClient(Input PML-File){
43      //PLM file is delivered to the client
44  }
45
46  end class
47
48
49  main (){
50          PROSTEP.ExtractionPointCloudFromTheExchangeMemory(Input PointCloudProcessPlant)
51          PROSTEP.Segmentation(Input PointCloudProcessPlant)
52          PROSTEP.Clustering(Input SegmentatedPointCloud)
53          PROSTEP.GeometryInformation(Input ClusterFile)
54          PROSTEP.CreationXML(Input CSV-File)
55          CooperationPartner.ExtractionCAD-ModelAndPIDfromExchangeMemory(Input CAD-Model, PID)
56          CooperationPartner.LinkingXMlwithPID(Input XML-File, CAD-Model, PID)
57          CooperationPartner.CreationPML(Input LinkedXML-File)
58          PROSTEP.DeliveryClient(Input ClusterFile)
59          CooperationPartner.DeliveryClient(Input PML-File)
60  }
```

Fig. 2.8 Pseudocode verification cooperation process

The corresponding inputs of the process steps are specified behind the singular functions. The comments written in green briefly describe what is done in the respective function. After "return" the specific output of the corresponding process step is indicated: the point cloud of the process plant, the segmented point cloud, the cluster file, and the Geometry File. In line 27, the class PROSTEP is closed.

The class CooperationPartner is opened in line 29 and the process steps which run on the premise of the cooperation partner are listed to the same extent. The process steps associated with the partner are the following: the extraction of the CAD model and the P&ID from the exchange memory, the linking of the geometry file with P&ID, the creation of piping modeling language (PML), and the final delivery of both the linked geometry file and PML file to the client. Afterward, in line 47, the class CooperationPartner is closed.

In line 50, the main process starts which collects and summarizes the singular 10 steps on the PROSTEP and the partner premise. The corresponding process steps of the cooperation process are seen in the curly brackets: in total 6 on PROSTEP premise, and 4 on partner premise. Before the process steps, the executing instance is written.

The pseudo-code illustrates the respective process steps with the corresponding inputs, the boundary conditions, and the outputs in line with the overall modeling progress (see Sect. 2.3.1 "Cooperation process"). The process steps start one after the other so that the cumulated cooperation process with two interfaces has been verified.

After the cooperation process has been verified, the internal process is verified in the same way in the following. Here, too, the process is illustrated by a pseudo code. The principle of the pseudo-code of the internal process is the same as that of the cooperation process but implemented in more detail with additional optimization and quality control steps which help to preserve the consistency of the entire process and the output data.

The verification of the internal process is shown in Fig. 2.9. First, the PROSTEP class is open. Then the process steps with the corresponding input and output values are specified with quality control (e.g., check and optimization of the geometry file). After the functions of the PROSTEP and the cooperation partner have been specified as already described in the cooperation process, the main process is started. For this purpose, the corresponding process steps are listed with the corresponding executing instances. The internal process can thus also be recognized as verified.

The verification lays the foundation for the subsequent validation of the concept based on a case study. For this purpose, the pseudo-code approach provides various options for the impact analysis [29]: stop the process, return or repeat one or a sequence of steps, handling errors.

```
begin class PROSTEP

function TotalSegmentation(Input PointCloudOfThePlant){
    //Segmentation of the point cloud
    return SegmentatedPointCloud
}
function SegmentationPipingParts (Input SegmentatedPointCloud){
    //Segmentation point cloud piping parts
    return SegmentatedPipeComponents
}
function Clustering (Input SegmentatedPipeComponents){
    //Clustering of the pipe components
    return ClusterFile
}
function RecognitionGeometryPipingParts (Input ClusterFile){
    //Recognition of the geometry of the piping parts
    return GeometryInformation
}
function CSV-Export (Input GeometryInformation){
    //Conversion of the cluster file into a CSV file
    return CSV-File
}
function OptimizationCSV-File (Input CSV-File){
    //Customization of the CSV file
    return OptimizedCSV-File
}
function CreationXML-File (Input OptimizedCSV-File){
    //Conversion of the CSV file into an XML file
    return XML-File
}
function DeliveryCooperationPartner (Input XML-File){
    //Delivery of the XML file to the cooperation partner
    return PML-File
}
function DeliveryClient (Input ClusterFile){
    //Delivery of the cluster file to the client
}

end class PROSTEP

begin class CooperationPartner

function DeliveryClient (Input PML-File){
    //Delivery of the PLM file to the client
}

end class CooperationPartner

main (){
        PROSTEP.TotalSegmentation(Input PointCloudOfThePlant)
        PROSTEP.SegmentationPipingParts (Input SegmentatedPointCloud)
        PROSTEP.Clustering (Input SegmentatedPipeComponents)
        PROSTEP.RecognitionGeometryPipingParts (Input ClusterFile)
        PROSTEP.CSV-Export (Input GeometryInformation)
        PROSTEP.OptimizationCSV-File(Input CSV-File)
        PROSTEP.CreationXML-File(Input OptimizedCSV-File)
        PROSTEP.DeliveryCooperationPartner(Input XML-File)
        PROSTEP.DeliveryClient(Input ClusterFile)
        CooperationPartner.DeliveryClient(Input PML-File)
}
```

Fig. 2.9 Pseudocode verification internal process

2.4 Conclusions and Outlook

The design, manufacturing, operation, through-life services, and support of modern process plants can be handled effectively using a concurrent engineering approach that focuses on the plant as a system and plant development lifecycle modeling. This requires a deep digitalization of processes and methods. These engineering systems are working in an environment that has multiple individual users and complicated supply chains. Additionally, their performance is affected by many other stakeholders. For both greenfield and brownfield plants, a digital representation of their assets and an individual set of rules are necessary in order to operate with different architectures and system processes during the entire lifecycle [30].

The stakeholders from different domains needed for the realization of complex systems typically have their own specific development methodology, standards, and terminology. Requirements Engineering needs to support the translation of requirements between the domains to enable a common understanding of the desired system [2]. This chapter describes the transition of a requirements model, represented by a structure of functional and non-functional requirements in a matrix, to a flexible, multidimensional process model modeled in the Cameo Systems Modeler. The system model is based on a consistent data model with predefined and optional interfaces to be able to link more data for further purposes.

Within the methodology, a model to describe interdependencies between the tangible and intangible components as well as between the main contractor and subcontractor was implemented. In the context of the digital twin, this model provides several advantages: many different data types can be linked to each other in a consistent way. If one element is changed, the change is reflected throughout the model. While the process model can be refined, updated, and improved, the presentation of it with traceable connections provides an overview of these complex interrelationships. Additional information can be linked to the model elements with the unique set of boundary conditions. It is possible to generate different views of the system model with a different focus and validation [29].

Further integration of system components from different domains will yield the collaboration and interoperation between previously separated domains like mechanical, chemical, and IT [31]. System operations are embedded in business networks that are continuously evolving and changing all the time. Since the system agents voluntarily participate in the network, they can come and go at any time without warning. This highly unpredictable relationship requires a different modeling approach [32].

New methodologies need to be developed to support interoperability between system components from different domains and describe the emergent system behavior [33]. This includes the management of unstable and unknowable requirements, taking into account information from all system lifecycle phases. Moreover, the value chains must be configured so that the system can be adapted to changing requirements through lifecycle services, even in the operation phase [34].

References

1. N.N. (1990) IEEE 610.12-1990, IEEE Standard Glossary of Software Engineering Terminology, IEEE, C/S2ESC—Software & Systems Engineering Standards Committee. https://standards.ieee.org/ieee/610.12/855/. Accessed 21 July 2022
2. Wiesner S, Hauge JB, Thoben KD, Peruzzini M (2015) Requirements engineering. In: Stjepandic J, Wognum PM, Verhangen WJC (eds) Concurrent engineering in the 21st century: foundations, developments and challenges. Springer International, Cham, pp 103–132. https://doi.org/10.1007/978-3-319-13776-6
3. Nuseibeh B, Easterbrook S (2000) Requirements engineering: a roadmap. In: Proceedings of the conference on the future of software engineering, Limerick. https://doi.org/10.1145/336512.336523
4. Stolt R, André S (2022) Requirements handling in multidisciplinary product development—a company study. Advances in transdisciplinary engineering, Vol. 28, 115–124. https://doi.org/10.3233/ATDE220638
5. Hauksdóttir D, Mortensen NH, Nielsen PE (2013) Identification of a reusable requirements structure for embedded products in a dynamic market environment. Comput Ind 64(4):351–362. https://doi.org/10.1016/j.compind.2012.10.008
6. Boehm B, Basili B (2001) Software defect reduction top 10 list. IEEE Comput 34(1):135–137. https://doi.org/10.1109/2.962984
7. Fougeres AJ et al (2022) A transdisciplinary method for requirements and BOMs engineering and their chaining toward CAD models. In: Advances in transdisciplinary engineering, Vol. 28, 93–102. https://doi.org/10.3233/ATDE220635
8. Adedjouma M (2012) Requirements engineering process according to automotive standards in a model-driven framework. PhD thesis. Université Paris Sud—Paris XI
9. Stark J (2015) Product lifecycle management, vol 1. Springer International Publishing, Switzerland. https://doi.org/10.1007/978-3-319-17440-2
10. Szejka AL, Aubry A, Panetto H, Canciglieri Jr O, Rocha Loures E (2015) A method for formalizing requirements interoperation in complex systems engineering. INSIGHT—Int Council Syst Eng (INCOSE) 18(4): 28–30. https://doi.org/10.1002/inst.12058
11. Hull E, Jackson K, Dick J (2011) Requirements engineering, 3rd edn. Springer-Verlag, London. https://doi.org/10.1007/978-3-319-61073-3
12. Nilsson S, Buffoni L, Sandahl K, Johansson H, Tahir Sheikh B (2018) Empirical study of requirements engineering in cross domain development. In: Proceedings of the DESIGN 2018 15th international design conference. pp 857–868. https://doi.org/10.21278/idc.2018.0466
13. NN The design structure matrix: introduction to the matrix. https://dsmweb.org/introduction-to-dsm/. Accessed 10 Jan 2022
14. NN (2022) Cameo systems modeler. https://www.3ds.com/products-services/catia/products/no-magic/cameo-systems-modeler/. Accessed 10 Jan 2023
15. Anderl R (2021) Vernetzte Produktentstehungsprozesse. TU Darmstadt, Darmstadt
16. Arndt D, Melzer S, God R, Sieber M (2017) Konzept zur Verhaltensmodellierung mit der Systems Modeling Language (SysML) zur Simulation varianten Systemverhaltens. Tag des Systems Engineering. https://doi.org/10.3139/9783446455467.012
17. Piaszczyk C (2011) Model based systems engineering with department of defense architectural framework. Syst Eng 14(3):305–326. https://doi.org/10.1002/sys.20180
18. Hallqvist J, Larsson J (2016) Introducing MBSE By using systems engineering principles. INCOSE Int Symp 26(1):512–525. https://doi.org/10.1002/j.2334-5837.2016.00175.x
19. Stjepandić J, Sommer M, Denkena B (2022) DigiTwin: an approach for production process optimization in a built environment. Springer International Publishing, Switzerland. https://doi.org/10.1007/978-3-030-77539-1
20. Stjepandić J, Sommer M (2022) Object recognition methods in a built environment. In: Stjepandić J et al (eds) DigiTwin: an approach for production process optimization in a built environment. Springer International Publishing, Switzerland, pp 103–134. https://doi.org/10.1007/978-3-030-77539-1_6

21. Stjepandić J, Bondar S, Korol W (2022) Object recognition findings in a built environment. In: Stjepandić J et al (eds) DigiTwin: an approach for production process optimization in a built environment. Springer International Publishing, Switzerland, pp 155–180. https://doi.org/10.1007/978-3-030-77539-1_8
22. Giliyana S, Salonen A, Bengtsson M (2022) Perspectives on smart maintenance technologies—a case study in large manufacturing companies. Adv Transdisciplinary Eng 21:255–266. https://doi.org/10.3233/ATDE220145
23. Biahmou A, Stjepandić J (2016) Towards agile enterprise rights management in engineering collaboration. Int J Agile Syst Manage 9(4):302–325. https://doi.org/10.1504/IJASM.2016.081564
24. Stjepandić J, Sommer M, Stobrawa S (2022) The commercialization of digital twin by an extension of a business ecosystem. In: Stjepandić J et al (eds) DigiTwin: an approach for production process optimization in a built environment. Springer International Publishing, Switzerland, pp 205–234. https://doi.org/10.1007/978-3-030-77539-1_10
25. Hanna M, Schwede LN, Krause D (2018) Model-based consistency for design for variety and modularization. In: 20th international dependency and structure modeling conference, DSM 2018, Trieste, Italy, October 15–17, pp 239–248
26. Sommer M, Stjepandić J, Stobrawa S (2021) Incremental update of a digital twin of a production system by using scan and object recognition. Adv Transdisciplinary Eng 16:83–92. https://doi.org/10.3233/ATDE210086
27. Peltsverger S, Debnath S (2019) Instructional pseudocode guide to teach problem-solving. In: ITiCSE '19: proceedings of the 2019 ACM conference on innovation and technology in computer science education. https://doi.org/10.1145/3304221.3325581
28. Elgh F, Johansson J (2019) Traceability in engineer-to-order businesses. In: Stjepandić J et al (eds) Systems engineering in research and industrial practice. Springer Nature, Switzerland, pp 115–146. https://doi.org/10.1007/978-3-030-33312-6_5
29. Schwede LN, Hanna M, Wortmann N, Krause D (2019) Consistent modelling of the impact model of modular product structures with linking boundary conditions in SysML. In: Proceedings of the 22nd International Conference on Engineering Design (ICED19), Delft, The Netherlands, 5–8 Aug 2019. https://doi.org/10.1017/dsi.2019.367
30. Mo JPT, Beckett RC (2019). System of systems modelling. In: Stjepandić J, Wognum N, Verhagen WJC (eds) Systems engineering in research and industrial practice. Springer, Cham. https://doi.org/10.1007/978-3-030-33312-6_4
31. Bicocchi N, Cabri G, Mandreoli F, Mecella M (2018) Dealing with data and software interoperability issues in digital factories. Adv Transdisciplinary Eng 7:13–22. https://doi.org/10.3233/978-1-61499-898-3-13
32. Hsu J (2019) Decision analysis and interface management in systems engineering. In: Stjepandic J et al (eds) Systems engineering in research and industrial practice. Springer Nature, Switzerland, pp 147–166. https://doi.org/10.1007/978-3-030-33312-6_6
33. Hsu J (2019) Fundamentals of systems engineering—a practitioner's approach. In: Stjepandić J et al (eds) Systems engineering in research and industrial practice. Springer Nature, Switzerland, pp. 19–52. https://doi.org/10.1007/978-3-030-33312-6_2
34. Andre S, Stolt R, Elgh F, Johansson J, Poorkiany M (2014) Managing fluctuating requirements by platforms defined in the interface between technology and product development. Adv Transdisciplinary Eng 424–433. https://doi.org/10.3233/978-1-61499-440-4-424

Chapter 3
Literature Review to Digital Twin of a Process Plant

Abstract The current state of the art of the digital twin is portrayed in this chapter with a focus on the process plant. For this purpose, an exhaustive sample of recent scientific literature was analyzed, and different approaches are explained. While it should be noted that a variety of definitions of the digital twin exist, several definitions of the digital twin are compared here. A digital twin is understood as a digital representation of an active, unique product or unique product-service-system with its selected characteristics during certain lifecycle phases. Countless publications document comprehensive approaches, advanced methods, and convincing practical solutions on how products, processes, and services can be optimized, and new ones created by using a digital twin. A taxonomy of the digital twin is elaborated. The multitude of practical applications of the digital twin can be structured according to the digital twin paradigm: functions, components, lifecycle, architecture, and context. An overview of the main application fields of the digital twin in different industrial domains is also drawn. The published applications of digital twins in the process industry are elaborated. The applications in the process industry are explored for use cases such as the design of digital twins, the process synthesis, the modernization of existing plants, incident studies, the planning of maintenance measures, the virtual monitoring of plant operation, and the training of operating and maintenance personnel. In the discussion section, the match is presented on how the digital twin provides the suitable approaches to business challenges a modern company in the process industry is faced with by timely sharing information about the products, processes, and resources across the lifecycle. The benefits of the digital twin are highlighted such as acceleration of business tasks and processes such as analytics, decision support, forecast/prediction, and recommendations. The distinction between the data-based and the system-based approach for the generation of the digital twin is elaborated. The gap in coherence between the real and the digital twin in a brownfield environment was detected. Finally, based on the digital twin characteristics, an approach for the generation and update of the digital twin of a process plant is drafted which should be offered as a commercial service.

Keywords Digital Twin · Product Lifecycle Management · Plant engineering · Process plant · Brownfield plant · Piping & Instrumentation Diagram

J. Stjepandić et al., *Generation and Update of a Digital Twin in a Process Plant*,
https://doi.org/10.1007/978-3-031-47316-6_3

3.1 Introduction

Enterprises of every magnitude have been confronted with faster, more complex, or more insecure boundary conditions over the past years. One of the biggest drivers of this phenomenon is digitization, entering more and more into every company's everyday life [1, 2]. Thus, companies are on the one hand forced to constantly develop or improve their processes, on the other hand, digital technology makes it possible to handle those challenges [2, 3]. During this journey, the approach of the digital twin has emerged over the past decades [4]. The term itself fundamentally describes a virtual representation of a physical system. A large amount of literature about the digital twin exists, nevertheless, there is not yet a consistent, widely accepted definition [5].

The term 'digital twin' has become popular in both business and academia recently. Countless publications document comprehensive approaches, advanced methods, and convincing practical solutions on how products, processes, and services can be optimized, and new ones created by using a digital twin [6]. In industrial usage, the digital twin has been recognized and adopted both by user companies and software vendors as a powerful means and the basis for further developments [7]. While the digital twin originates from the manufacturing domain, its recent underpinning technology maturation makes it suitable for all those domains where there is a need for studying virtual interactions with the physical environment [8]. The current state of the art of the digital twin is portrayed in this chapter with particular respect to process plants. A comprehensive introduction to the digital twin and the underlying solution approach was given in the previous book [9].

The virtual environment provides good capabilities to simulate and analyze the content that is available in the virtual environment [10]. Thus, the interoperation of the analyses and simulations should consider how up-to-date the virtual environment is [11]. To have the information (e.g., spatial positions) put from the physical equipment to the virtual environment requires higher precision. Conversely, the virtual environment is as accurate as needed and the trust in the correctness of the virtual environment is conserved [12, 13].

As mentioned in the introduction, the generation and the update of the digital twin for an existing process plant lies in focus [14, 15]. For this purpose, different approaches from today's literature are investigated, and different definitions of the digital twin are depicted [16]. At this point, it should be noted that a variety of definitions of the digital twin exist [17]. That is why only a selection of literature, providing a good introduction to the subject, is described in this chapter. After defining the digital twin, its three particular aspects are described more deeply.

In this chapter, the essential achievements of the digital twin are presented. In Sect. 3.2, the definition and taxonomy of the digital twin are briefly discussed together with its achievement as recorded in the literature. In Sect. 3.3, the main applications of the digital twin are presented, followed in Sect. 3.4 by applications of the digital twin in the process plant. In Sect. 3.5, the achievements of the digital twin are discussed. The chapter ends with a summary and future challenges.

3.2 Definition and Taxonomy of Digital Twin

The subject of the digital twin was presented for the first time at the University of Michigan in 2003 by Michael Grieves. This lecture referred to Product Lifecycle Management (PLM), but the basic characteristics have not changed ever since [18]. M. Grieves defined a "real" and a "virtual" space, linked through an exchange of data and information. In 2010 the basic concept of the digital twin was incorporated into a NASA-project, by simulating the behavior of a spaceship capsule in space. In that context today's most common definition of a digital twin emerged, referring to every phase of a lifecycle [19]:

> *"The digital twin is an integrated multi-physical, multi-scale, probabilistic simulation of a vehicle or a system which uses the most advanced physical models, sensor updates, fleet history, etc. available, to mirror the life of its flying twin. The digital twin is extremely realistic and can consider several important and interdependent vehicle systems."*

With the rise of IT technologies such as Cloud Computing, the Internet of Things (IoT), Big Data, and Machine Learning (ML) some significant changes took place in the industrial areas, for example, Industry 4.0 [20]. Furthermore, the rise of artificial intelligence has a particular influence on the rapid dissemination of the digital twin [21] by facilitating new model-based representations and solution approaches [22]. Traditional model-based approaches are being adapted to the evolving boundary conditions and provide a demand-oriented, real-time capable evaluation basis to efficiently support decision-making in multi-objective problems [23].

Following this technological progress, the manufacturing industry has used the digital twin in different areas for a variety of tasks such as quality assessment, predictive maintenance, and shop-floor monitoring [24, 25]. Today, the digital twin is understood as a subject of scientific research, as well as a keyword to advertise existing products and services from a new point of view [25, 26]. However, the implementation of the digital twin is still considered as a company-specific and complex procedure, because the technical and organizational options vary heavily [27].

Definitions of the digital twin changed over time until in 2017 Stark et al. came out with a definition for the digital twin [28]:

> *"A digital twin is a representation of a unique asset (product, machine, service, product-service system or a different immaterial asset) that represents its characteristics, condition, and behavior using models, information and data."*

In the context of this approach, the definition of Stark et al. is taken as a basis [28]. Following that definition, the digital twin can exist throughout one single or several lifecycle phases.

The digital twin obtained special attention during the rise of digitization in all aspects of life and the increased usage of cyber-physical systems and cyber-physical product systems [29]. In particular, the combination of Industry 4.0 and artificial intelligence provides a fertile ground for the exploitation of the digital twin [23]. All entities in a digitalized world can collect, create and process data and information [30]. Due to that feature a digital shadow arises, which can be used to depict a system's condition [31]. In addition to that, an assessment database is formed over time that assists with describing the behavior of a system [32]. By integrating this digital shadow of a product into general models, a digital twin can be created [33]. These models are saved in a digital master [34] or derived from a digital prototype (e.g., computer-aided design model) [35] or a simulation model (e.g., computer-aided construction model) [36].

In their publication [37], Kritzinger et al. provide an overview of how to distinguish between the concept of the digital model, the digital shadow, and the digital twin. According to that, a digital model is defined as a "digital representation of an existing or planned physical object. There is no automatic data exchange between the physical and the digital object". A model can contain a description of a physical object. For example, a digital model can be a simulation model of a planned factory, a mathematical model of new products, or different models of a physical product, that do not use any characteristic of automatic data integration. The digital data still can be used however data exchange only occurs on behalf of human interaction. Changes in the digital or physical object are not synchronized automatically (no automatic data flow) and have no effects on the respective other. The following Fig. 3.1 illustrates the data exchange of a digital model.

A digital shadow, on the other hand, can be understood as a one-way automated data flow from the physical to the digital object based on the definition of the model [38]. The change of condition of a physical object leads to a change in its digital shadow, but not the other way around. In a wider context, it is understood as a platform that integrates information from different sources in order to enable miscellaneous real-time analyses for decision-making.

For a better understanding of the data flow, it is illustrated in the following Fig. 3.2.

A digital twin requires and enables data flow in both directions. Furthermore, the data flow is an inherent constituent of the digital twin. Therefore, the digital object can be considered as a control entity for the physical object to replicate processes in order to collect data to predict how they will perform. Changes to either the physical or the digital object simultaneously cause a change in the respective other object. The frequency and extent of the data exchange determine the performance of the digital twin. The data flow of a digital twin is illustrated in the following Fig. 3.3.

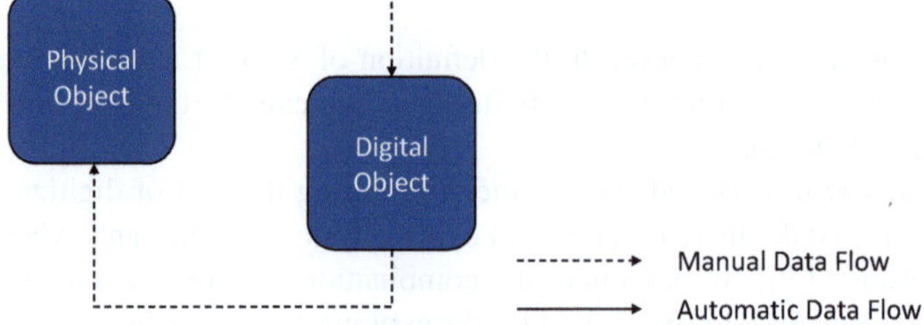

Fig. 3.1 Data exchange of a digital model

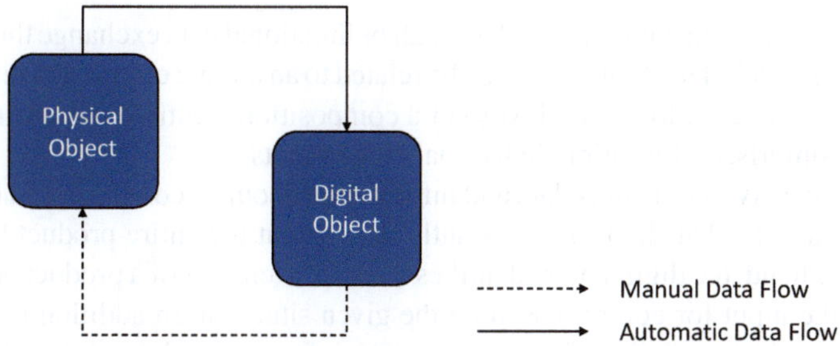

Fig. 3.2 Data exchange of a digital shadow

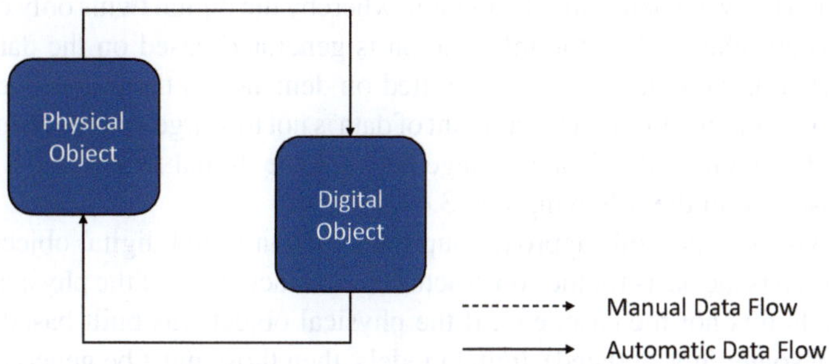

Fig. 3.3 Data exchange of a digital twin

Based on the findings so far, the digital twin consists after Stark et al. of three constituents [28]:

1. A unique instance of the digital master model (and/or digital prototype) that is tailored to the specific purpose.
2. An individual digital shadow of a product which means, that said digital shadow consists of data, collected or measured throughout the operation and usage of the product, the device, or the machine.
3. A useful connection between the digital master (and/or the digital prototype) and the digital shadow.

Trauer et al. build on these properties in the literature and defines general characteristics that should describe a digital twin [6]:

1. The digital twin is a virtual and dynamic representation of a physical artifact or system.
2. Data is automatically and bidirectionally exchanged between the digital twin and the physical object.
3. The digital twin contains data from all lifecycle phases and is connected to all phases.

Based on these characteristics, a general definition was derived that is widely used in the industry. According to that, a digital twin is "a virtual dynamic representation of a physical system which is connected through bidirectional data exchange throughout its entire lifecycle". The digital twin can be related to an atomic entity (discrete digital twin e.g., a product of low complexity) or a compositional entity (composite digital twin e.g., comprised of multiple individual components).

The physical twin transmits data and information about its condition and behavior to the digital twin. This happens constantly throughout the entire product lifecycle. On the other hand, the digital twin identifies the improvements of a product or process and takes the input for control based on the given situation. In addition to that, the digital twin makes predictions for the future. The information about the predictions is directly transmitted to the physical twin so that it can make appropriate adjustments.

As already described, data exchange takes place fully automatically. It is to be distinguished between data and information, whereby the digital twins only exchange data with each other. While the information is generated based on the data in real space afterward, the data is only transmitted on demand, so there is no permanent transmission. This ensures that the amount of data is not too large to be exchanged. For a better understanding, the data exchange between the digital twin and the physical twin is illustrated in the following Fig. 3.4.

As we will see later, this approach supposes that an initial digital object (digital model), which is the basis for the construction (manufacturing) of the physical object, still exists. Is this not the case, e.g., if the physical object was built based on non-digital or obsolete (not-updated) digital models, then those must be generated based on the characteristics of the physical objects. Due to its long lifecycles, this is a frequent case in the plant industry.

In Fig. 3.4 it can be seen that the physical twin is connected to the digital twin throughout every phase of the product lifecycle. The physical twin is located in real space sending data about its behavior to the digital twin in the virtual space. It receives the data accordingly and assigns it to the responsible digital twin. Different

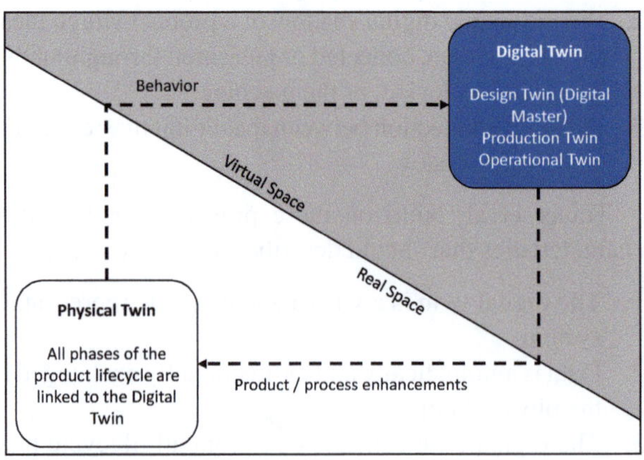

Fig. 3.4 Data exchange between the digital twin and the physical twin

digital twins can exist from one physical twin, all with different data. They are represented in the graphic by the production twin and the operational twin. All identified enhancements of the product or the process are transmitted to the physical twin. This leads to continuous data exchange between the digital and physical twins.

The data flow between the physical and virtual twin ensures that the digital twin can be operated continuously. The connection between the components is necessary to allow the interaction between the elements of the digital twin. Following Adamenko et al., these connections can be separated into three groups [39]:

- Connections in physical space
- Connections in virtual space
- Connections between virtual and physical space.

Finally, we will use the taxonomy here to highlight the key distinguishing features and properties of digital twins. Among various sources, a compact approach by van der Valk et al. was selected which comprises three non-mutually exclusive dimensions and five mutually exclusive dimensions, as presented in Table 3.1 [40]. Each dimension has two or three characteristics that look like binary.

The dimension **data link** assesses the communication between digital and physical twins, which can either be one-directional or bi-directional, and makes this dimension mutually exclusive [40].

With regards to the **purpose** of the digital twin, it is determined by the way of data handling: processing data (1), transferring data (2) from one point (the physical

Table 3.1 Taxonomy of digital twin, derived from [40]

Dimension	Characteristics	Exclusivity
Data link	One-directional Bi-directional	Mutual
Purpose	Processing Transfer Repository	Not
Conceptual elements	Physically independent Physically bound	Mutual
Model accuracy	Identical Partial	Mutual
Interface	M2M HMI	Not
Synchronization	With Without	Mutual
Data input	Raw data Processed data	Not
Time of creation	Physical part first Digital part first Simultaneously	Mutual

part) to another one (data warehouse), and data repository (3). This dimension is not mutually exclusive because the digital twin may have one, two, or all three characteristics at the same time [40].

Conceptual elements show how deeply the digital and physical twin are integrated, if the digital twin can be directly bound to its physical twin in a one-to-one ratio, or if it is independent. This dimension is mutually exclusive [40].

Model accuracy expresses whether the digital and physical twin work with identical accuracy or partial accuracy (e.g. when a digital image only reflects crucial parts of the physical object). The model accuracy is mutually exclusive [40].

Multiple choices are possible at the dimension **interface**: a human–machine interface, a machine-to-machine interface, or both, a human–machine interface and a machine-to-machine interface. This dimension is not mutually exclusive [40].

Synchronization reflects the working synchronization between the digital and physical twin by (real-time) data updates during its lifecycle or without synchronization at all. The synchronization is mutually exclusive [40].

Data input differentiates between pure, raw data gathered directly from sensors or data, which is pre-processed (e.g., by analytic software) before it is transferred to the digital twins. Data input is not mutually exclusive [40].

The **time of creation** describes the chronological order in which the respective parts of the digital twin come into existence. Thus, the dimension distinguishes whether the physical part or the digital part is developed first, or both parts are developed simultaneously. Most digital twins are designed after a physical system [40].

3.3 Main Applications of the Digital Twin

The multitude of practical applications of the digital twin can be structured according to the digital twin paradigm which provides answers to questions depicted in Fig. 3.5 [41]:

1. Functions (Why?)
2. Components (How?)
3. Lifecycle (When?)
4. Architecture (How?)
5. Context (Where?).

However, the architecture and the components can be summarized as the technology, the functions, and the lifecycle comprise the scope. The context belongs to the domain of exploitation.

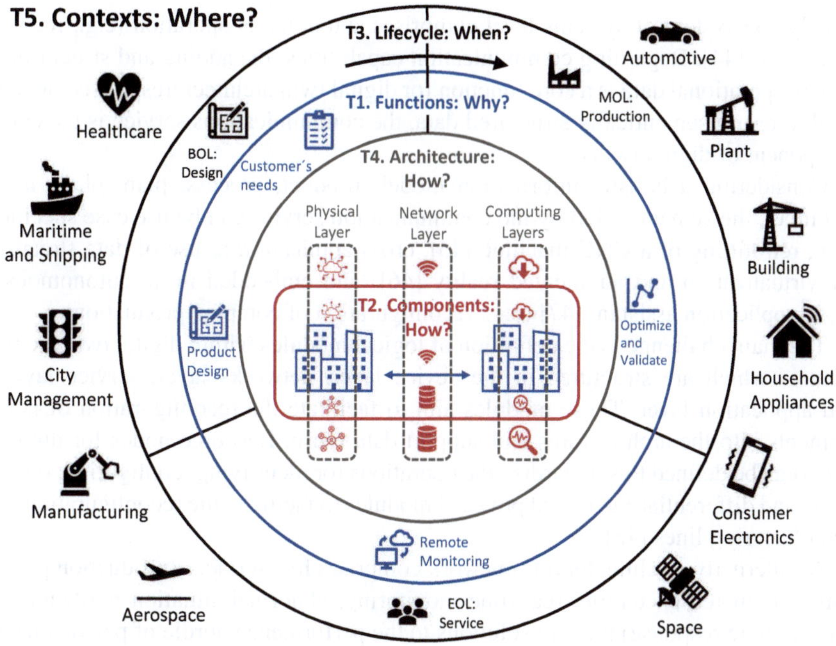

Fig. 3.5 The digital twin paradigm, derived from [41]

3.3.1 Digital Twin Technologies

The digital twin consists of a set of models with complex structures and behavior, that mimic the real-time operations of the physical system. A digital twin can be a surrogate of a component, a system of components, or a system of systems. Digital twin architectures are determined by a scalable synchronization of data from a physical twin to a virtual model via a communication service with accurate models of reality. Based on 19 themes in total needed to characterize the digital twin, the technology comprises 6 themes [42]. Digital twins can be shaped according to different types, which are covered by the scope. The types respectively distinguish between unit level (individual products, machines), systems level (complex products, productions lines), and system of systems level (complete plants and companies) The physical component's scope distinguishes all types of digital twins. Those types cover the unit level (individual products, machines), the systems level (complex products, production lines) up to the system of systems level (complete plants and companies) [43].

The lack of a (quasi) standard reference architecture yields the development of digital twin derivates based on a plethora of existing components, interfaces, communication protocols, models, and data [41]. On the unit level, the focus is set on improvements in product/machine behavior and continuous operation. Further capabilities including planning and self-optimization are covered at the systems level.

Finally, the system of systems level comprises autonomous operation (e.g., reconfiguration) [44]. Regarding communication capabilities, the adding and structuring of new operational data is a core function for digital twin architectures. Based on the need to re-use semantically structured data, the communications service is the core component of digital twins.

Considering a holistic information model (product, process, plan, plant, and resource), the semantics fed by the communication service can be use case specific (e.g., retrofitting of a CNC machine [45], cross-application re-use of data streams for virtualization and augmented reality [46]) and embedded in an autonomous, cross-application assistant [47] (e.g., reconfiguration of complex production).

The main challenge is the definition of logical modules within digital twin architectures which are structured in the device layer, networks layer, service layer, and application layer. These modules aim to facilitate the reconfiguration of new elements into the architectures and support data reuse. Resource nodes for digital twins can be defined to standardize the operations for identifying, configuring, operating, and differentiating logical physical modules to facilitate the reconfiguration of manufacturing lines [48].

An alternative architectural framework comprises five services (production planning, automated execution, real-time monitoring, abnormal situation notification, and dynamic response) that are solutions to the performance hurdle of personalized production. This framework operates on information from the proposed product, process, plan, plant, and resource (P4R) information model [49]. The reference architecture of related technologies such as Industry 4.0 can be partially reused [50].

The ongoing and further development of the digital twin technology is being directed to gather the following improvements: (1) How to model physical objects into the virtual objects underlying the digital twin concept; (2) how to structure the architecture of the digital twin; (3) how to synchronize real-time data between both physical and virtual components of the digital twin based on a secure connection [51]. The concept presented here provides one implementation of the direction (1).

3.3.2 Digital Twin Scope

While the digital twin refers to the entire product lifecycle, the related phase can be used as the criterion for the scope: design and development, production, and service [5]. Furthermore, the digital twin is often extended to cover the operational dynamics in the supply chain throughout all phases of the product life cycle [52]. Digital twin applications are mainly developed for prediction purposes and used for decision-making support. However, most digital twin applications refer to a single phase of a product life cycle [41]. During the phase of product design and development, the digital twin can be applied in conceptual design, detailed design, design verification, and redesign [5].

Within the conceptual design, the designer should consider a large number of historical data, information, and knowledge from the market, customer, manufacturing, suppliers, and internal parties [5]. Using the data and information collected from the product Middle of Life (MOL) can facilitate the emergence of a new design concept [53]. In the detailed design phase, there are involved many more factors and many parties. Subsequently, the digital twin can benefit from the connection to existing data sources, and the data management approach is used that improve the data monitoring and exchange in the product lifecycle [54]. The outcome can be the derivation of digital twins for availability-oriented business models [55]. A plethora of applications are dedicated to verification and validation. The visualization capability and collaboration of different parties can be improved using virtual and augmented reality [56].

Production is the most frequent application field of the digital twin [57]. The digital twin provides a virtual design method to provide a geometric query between virtual process twins that allow users to generate virtual processes with resource reference in early production engineering design phases, without having to know every single detail about product structures [41]. Further applications are in production planning and resource allocation [58], production process simulation (e.g., assembly, etc.) [59], optimization of the industrial human–robot collaboration [60], enabling sustainability (e.g., energy control, waste control) [61], enabling Industry 4.0 and cyber-physical production system [62], production tool or machine monitoring and prognosis [63], and optimization of warehouse management [64].

The digital twin in service comprises components that handle the reception, formatting, and processing of the operational state data, provide a digital model representing the salient properties, behavior, and operation of the physical twin over its twinned lifecycle, and deliver an interface enabling output from and interaction with the digital twin by humans or systems [53]. The digital twin aims to coordinate cross-organizational resources to provide on-demand services for personalized manufacturing requirements [65]. The digital twin improves the safety and reliability of operations through condition monitoring of the system in operation [66]. Conversely, a digital twin-driven service model is built to perform the seamless monitoring and control of shared product-service systems [67].

3.3.3 Digital Twin Industrial Domains

The existing digital twin applications refer to the specific and traditional contexts in almost all industrial domains: aerospace, space, automotive, maritime and shipping, manufacturing, household appliances, consumer electronics, healthcare, building, plant, transportation, and city management [41]. A close, deep interaction between the real and the virtual world with the involvement of humans as workers, operators, or users poses a promising and challenging target in all domains. Applications are mostly related to data-intensive tasks. Primarily, the applications can be distinguished

Table 3.2 Applications of digital twins in industrial domains, derived from [5]

Domain	Application	Reference
Aerospace	Virtual testing or simulation Diagnostics and prognostics of aircraft health	[68] [69]
Space	Operation and management for the entire service life of on-orbit spacecraft	[70]
Automotive	Digital testing and verification Driver intention prediction and automatic driving monitoring	[71] [72]
Maritime and shipping	Development of autonomous maritime surface vessels	[73]
Manufacturing	Operation monitoring and fault diagnosis Manufacturing system reconfiguration based on industrial human–robot collaboration	[74] [75]
Household appliances	Sustainable energy consumption	[76]
Consumer electronics	User experience of consumer haptic devices	[77]
Healthcare	Optimize elderly healthcare services Surgery or medical simulation	[78] [79]
Building	Building management and control Building health monitoring	[80] [81]
Plant	Optimize project planning and management Plant performance optimization and protection of the environment	[82] [83]
Transportation	Optimize traffic in an airport	[84]
City management	Sustainable smart city design and building information modeling	[85]

in the extent of application and the degree of maturity. A non-exhaustive excerpt of applications of digital twins in industrial domains is given in Table 3.2.

3.4 Application of Digital Twin in a Process Plant

The digital twin has been recognized as an attractive means to optimize plant engineering, in particular, related to their long lifecycles. Companies in the process industry must achieve competitive advantages, especially with existing (brownfield) plants whose optimal operation is made more difficult by using multiple, hardly connected systems with a low amount of shared data captured across different repositories. The digital twin tailored to these constraints must thus not only cope with the necessary integration of current and future information technologies and their applications in the process industry but in particular enable the gradual further development of brownfield plants with all their analog legacy to be designed in such a way that they have a positive long-term effect to efficiency [86]. The focus of the digital twin of a process plant can lie in the scalability of architecture, synthesis, simulation, optimization, safety, reconfiguration, and training for operation.

Digital twins are currently one of the preferred tools for improving the operational efficiency of the main and auxiliary equipment in a process plant. Nevertheless, the stakeholders from different domains have varying understandings and opinions about the digital twin. Most of the typical digital twin application scenarios belong to discrete manufacturing and are not suitable for the process industry with its multi-scale hierarchical and functional structure in space and time. Basically, the operational measurement information coming from the sensors should be compared with the results calculated in the digital twins of the equipment units and the output aims to detect the current technical condition and initiate the decision-making [87].

The combination of resource and process models to a capability model allows statements on the capability to perform process engineering operations with certain stationary and dynamic parameters and assured properties in existing or planned facilities. To represent the process hierarchy, the architecture of a software product was implemented for building a digital model of a thermal power plant thermal circuit by the allocation of layers. A three-level architectural solution of the digital twin construction subsystem has been developed including one abstract layer, the database management system layer. The main advantage of the method is scalability since thermal circuits can contain a large number of pieces of equipment. In such a way, a predictive analytics system is realized using the concept of the main and auxiliary equipment digital twins of energy facilities [87].

Process synthesis poses a further area of interest. For the specific case of fermentation operation, a central step in many biomanufacturing processes, a framework built upon a five-step pathway starting from a basic steady-state process model is proposed to develop a fully-fledged digital twin including the roles of both the engineers and the operators. The definition of the digital twin was extended by the Human Machine Interface (HMI) to make it applicable to fermentation operations. An effective HMI facilitates the communication between the staff and the digital model as both engineers and operators need to interact with it, the one reason being the physical operation interaction, the other being uncertainties in the measurements and the model. As a result, such an HMI should be designed so that the engineers are explicitly made aware of underlying assumptions and estimations of critical operational parameters and their ramifications. The ultimate aim is to ensure that engineers fact-check the outputs against their engineering judgment. The success or failure of a fully-fledged digital twin implementation is determined by key factors that comprise the role of modeling, human operator actions, and other propositions of economic value [88].

Simulation is an almost inexhaustible field for the application of the digital twin. One of the foremost challenging topics is the generation of appropriate high-fidelity simulation models based on standard procedures. A method for the automatic generation of a formal model of a plant was implemented from the behavior traces recorded from its digital twin. In this way, the formal verification of process models can be conducted with reduced effort. The traces are observed from simulation in the loop of the digital twin in Visual Components connected with distributed automation software, according to IEC 61499. The generated modular formal model of the closed-loop system was transformed into the model of uncontrolled plant behavior extended with non-determinism. The model is then combined in a closed loop with the formal model of the controller, generated from its source code. The verification and simulation were provided by the symbolic model checker, which verifies various specifications of the system [89]. Furthermore, time-consuming system integration is addressed by applying a lifecycle-wide online simulation architecture. This architecture also enables the possibility to apply model optimization and online estimation methods [90].

The optimization based on the digital twin requires dynamic problem-solving capabilities. Therefore, digital twins are widely used in performance improvement analysis in food processing companies despite limited implications for business strategies. The digital twin provides a benefit because it can perform simultaneous multithreaded work. During the operation of the food processing company, multiple factors must be considered at the same time to optimize various parameters simultaneously to solve a real-life problem. It demonstrates the various digital twin implementation stages, such as strategic mapping and physical-virtual space replica, with rigorous analysis. As a result, the physical-virtual interface model helps to increase the existing system's machine availability, allocation efficiency, technical efficiency, worker efficiency, utilization rate, effectiveness, step ratio, and throughput rate [91].

A wide field of applications poses the optimization of energy consumption. The energy transition in the process industries can be accelerated by process intensification and the digital twin. As a methodology for making remarkable reductions in equipment size, energy consumption, or waste generation while achieving a given production target, the process intensification utilizes the innovative principles in equipment design and control to improve the physical process, while the digital twin provides the virtual model of the plant as an environment for further optimization of the operational parameters. Hybrid methods attempt to combine the advantages of both heuristic as well as mathematical optimization methods. The effects of both tools on the energy transition are evaluated not only from the point of application but also from the possibility of implementation and limitations in process industries. Despite their benefits, the deployment of process intensification and the digital twin requires not only infrastructure and capital investment but the knowledge and collaboration of different levels of plant personnel [92].

Process safety is an important prerequisite in each process plant. While it is a subject of continuous assessment, it can be facilitated by a digital twin. The whole process of the petrochemical industry involves flammable and dangerous explosive goods. The timely discovery of abnormalities or failures is crucial to ensure production safety. After the process was decomposed into five levels one by one, the digital twin of a petrochemical process was built as a twin plug-in for each component of the component layers, and then inversely decouple the process to assemble the digital twins' layer by layer. Based on modules of temperature field, pressure field, and flow field, the logical structure of chemical process status monitoring and fault diagnosis in detail was composed, which improves the safety and controllability of the petrochemical process [93].

The acquisition of an existing plant, a so-called brownfield plant, poses a big challenge due to the high effort for manual work and potential errors. Conversely, the success of the retrofit or reconfiguration of existing plant systems heavily relies on knowledge about the system. Such work can be planned, supported, and simplified with the digital twin of the system providing this knowledge. However, digital models as the basis of a digital twin are usually missing for plants that were built long ago. A digital twin can be generated by extracting process data from engineering documents using text and image processing techniques based on a steady-state simulation model, using a Piping and Instrumentation Diagram (P&ID) as the main source of information. Methodology and toolchains are proposed, consisting of manual, semi-automated, and fully automated steps. A pilot-scale brownfield fiber processing plant was used as a case study to demonstrate the proposed methodology and toolchain and to identify and address issues that may not occur in laboratory-scale case studies [94].

Immersive technologies such as Augmented Reality (AR) and Virtual Reality (VR) are used in several tasks such as operator training or remote maintenance to gain benefits from high virtuality. The user study focuses on usability, user experience, degree of achievable sense of presence, and operator workload. In addition to improvements regarding task performance, benefits in user experience, immersion, and workload experienced are observed in all scales. As an example, operators such as Human-in-the-Loop are enabled to annotate new or not yet recognized objects as well as refine existing datasets for object recognition and pose estimation. In addition, imitation learning can contribute to successively increasing the degree of automation and minimizing required intervention incidents [95].

3.5 Discussion

As presented earlier in this chapter, digital twins already are broadly applied in the process industries, with the main intent of modeling a system, in line with the model-centric nature of engineering practice, and further emergence can be expected too [96]. First of all, the digital twin provides the appropriate approaches to many businesses challenges a modern company in the process industry is faced with (Table 3.3) by timely sharing information about the products, processes, and resources across the lifecycle. In particular, the digital twin accelerates business tasks and processes such as analytics, decision support, forecast/prediction, and recommendations.

The benefits of a digital twin are three-fold. First, virtual models driven by real-time data are capable to provide a representation with a high level of fidelity of the real environment and infrastructure and, as such, can be structured to support immersive interactions between humans and machines. Secondly, digital twin data can integrate, merge, and analyze real and simulated data, enabling the user to get a comprehensive description of the whole environment and a deeper understanding of

Table 3.3 Use of digital twin to mitigate challenges in the companies, derived from [66]

Challenge	Approach
Complexity	Increasing product, supply chain, and demand complexity mandates manufacturers to manage risk and safety, improve product and asset performance, and maintain high levels of enterprise-wide quality
Ecosystem	Extending and broadening external networks of suppliers and partners, many of whom provide design and operations support
Standardization	Highly competitive global markets need to be served at a local level with unique requirements, processes, methods, and capabilities
Customer experience	Closer collaboration with customers is needed for customized or individualized products across their life span
Data	Massive amounts of structured and unstructured data from IoT, extended supply chains, and multiple disparate manufacturing plants or facilities
Connected, "always-on"	Products and assets can be tracked throughout their life cycle, and customers expect high levels of quality and service
Trustworthiness	Providing a global view of security, reliability, and availability
Service revenue	Connectivity provides an opportunity for manufacturers to leverage connected assets and product data to ensure high levels of service, value-added services, and an increase in service revenue
Collaboration	Providing a shared understanding of what an entity looks like when it is represented in a digital format
Auditability	Enhancing the capability of a company to successfully pass an audit by regulatory bodies
Knowledge transfer	The "single source of truth" provides the potential for knowledge transfer to all stakeholders

the physical entities. Finally, digital models provide concise, intuitive, and easy-to-use platforms such as front-end software and mobile apps that could be adopted by users in different fields [96].

Building a digital twin requires modeling the static properties of the system: the system requirements and constraints, including the functions and functional decomposition, model data flow and communication, then the architecture and logical structure of the system [51]. Once the digital twin is built, the interactions between the elements of the digital twin provide a basis for continuous changes and optimizations. The question arises, of how a complex system such as a digital twin can be created in the first place. Therefore, Adamenko et al. present two different approaches in the literature, a data-based one and a system-based one [39]. There are further advanced approaches related to specific applications in the literature [51, 54, 55, 57] including the self-learning function [56]. In order to find the right approach, it is essential to know the specific features to be achieved with the digital twin.

The data-based approach puts the focus on data, structured by certain criteria. Thereby criteria can be e.g., different features or assemblies of a physical object. This provides a quick overview of the performance of an object. Sensors or other sources of data can be matched with the characteristics afterward. The data is then evaluated and analyzed with the help of algorithms and functions in combination with machine learning or similar methods. This enhances operative forecast models and foresight maintenance to be more precise. The data can be visualized for simplification in diagrams and other visualization options, linked to a plant model or stand-alone [97].

When using a data-based approach, the first thing to ensure is that the data is freely accessible. Then models, analyses, and functions are created, verified with data, and provided in the follow-up. Once this is done, the data is stored. An analysis of an object's behavior can take place over a long period. In order to enable the user to work with the data that is relevant to him, special front-end applications must be created. These applications can also be combined from several digital twins [56]. Physical and digital objects can be aligned bidirectionally so that both parts are always up to date. The data refers to the entire product lifecycle.

The data-based approach is an emergent process that incorporates the data over time. This means that not all data is always available and accessible. Likewise, decisions must be made not only about what data to collect, but also about how to collect data or how to use experiments. More of an advantage is that the data-based approach does not require the availability of all information to create a digital twin. Only the access to the sensor is necessary to analyze and evaluate the data. The structure of the data is independent of the aspired information of the user.

Unlike the data-based approach, the system-based approach centers around the physical object. Different models are combined to achieve the most accurate representation of reality possible. The system model represents the only source of truth and contains all information about the connections between the individual components. To create a system-based twin, all technical information (e.g., data sheets) or information about the software and the electronic parts must be available. For the creation of a system-based twin, a skeleton must be created which contains all information and can be enhanced with additional models and data. This skeleton is the foundation for the digital twin and is responsible for the data exchange between the different models and systems.

The model is best suited for a simulation, so that information, about how an object behaves under certain circumstances or changes, can be delivered. This is how certain situations can be simulated without interrupting the actual process. This allows for saving costs. In addition, rules can be created which can be used by the user in certain situations. The simulation model is able to learn because of the different settings so it can adjust those settings if a malfunction occurs. Since the most accurate model possible forms the basis of this approach, this approach is far more complicated than the data-based approach.

Nevertheless, the system-based approach allows a much more comprehensive statement to be made about the performance or capability of a plant. The simulations can not only be used to reveal existing failures but also to prevent any failures that may occur in the future. Different scenarios can thus be simulated and analyzed. The combination of the two approaches often is the best way to create a digital twin. Thus, the advantages of both approaches can be combined.

The aim of the approach presented in this book is to develop a system-based approach for the generation of an incremental digital twin of an existing process plant based on a 3D model of a scan. This incorporates the generation of a model of the entire plant as well as the portion of. A competing approach is not known at this time neither in the commercial publications nor in the scientific literature. In the literature, related approaches can be found for the recognition of building structure [98, 99], pipelines and their components [100–104], localization in space [105], and interpretation of the Piping & Instrumentation Diagram (P&ID) [106]. A roadmap to the semi-automatic generation of digital twins for brownfield process plants which provides a basis for further development and substantiation is proposed [107, 108]. All these approaches provide a solution for a singular problem but not for the entire plant. There is still a gap in research and development to generate a high-fidelity 3D model of an existing plant according to previous approaches of the authors [9].

Finally, there are several generic procedure models for the conception and implementation of a digital twin [109, 110] which are oriented to a software development process model. The relations between use cases, usage data, and virtual models resulted in a target concept as well as requirements for the implementation in five steps.

3.6 Conclusions and Outlook

As presented in this comprehensive literature review, the digital twin already provides profusely potential in various industries, in particular for maintenance, planning, scheduling, and control for the process industry [85]. Research and development in this field have emerged steadily in recent years and more and more applications can be found in practice [42, 57, 66, 111]. Conversely, the degree of penetration for the technology of the digital twin is still insufficient because there are so many plants in service that were built prior to the digital era and, thus, for which the digital models and documentation are not available yet [112]. In such cases, digitization should go hand in hand with modernization.

In this chapter, the different types of digital twins in different industries were first described. An important distinction was made according to taxonomy, technology, scope, and domain. Hence, the applications in the process industry were explored for use cases such as the design of digital twins, the process synthesis, the modernization of existing plants, incident studies, the planning of maintenance measures, the virtual monitoring of plant operation, and the training of operating and maintenance personnel [113–115].

While the digital twin is expected to integrate data from the lifetime of a product seamlessly, several key research and development challenges concern the definition of standards and communication protocols to ensure the interoperability of multiple digital twins with each other. There is a need for reference architecture with a unified framework to develop high-performance big data pipelines that allow for fast transfer, replication, and adoption of product engineering and manufacturing operational optimization [116].

One of the key aspects of digital twins is the generation and the update of them. The discussed use cases promise significant potential benefits, but also place different demands on the preparation and provision of digital information. The vision of coherence between the real and the digital twin even in a brownfield environment is being implemented more and more. Nevertheless, a significant gap in the research and development of commercial solutions was detected. It can be in general concluded that manual model generation and update is possible given the available methodologies, while guided and automated approaches still require substantial research and development to be conducted [117].

This book is dedicated to a flexible, cost-effective, and efficient approach with fast scans of the plant equipment and structure, object recognition, and a highly automated simulation model generation [9]. The further explanations in this book will go through the singular aspects of this approach which should be offered as a service for industrial customers. Intermediate results can be used for layout planning, maintenance planning, and documentation purposes. The service prepares the characteristics and the information from the digital twin in a way so that the user is able to access them without needing certain know-how [118].

References

1. Lee C-H, Trappey AJC, Liu C-L, Mo JPT, Desouza KC (2022) Design and management of digital transformations for value creation. Adv Eng Inform 52:101547. https://doi.org/10.1016/j.aei.2022.101547
2. Lee C-H, Liu C-L, Trappey AJC, Mo JPT, Desouza KC (2021) Understanding digital transformation in advanced manufacturing and engineering: a bibliometric analysis, topic modeling and research trend discovery. Adv Eng Inform 50:101428. https://doi.org/10.1016/j.aei.2021.101428
3. Wu Y, Zhou L, Zheng P, Sun Y, Zhang K (2022) A digital twin-based multidisciplinary collaborative design approach for complex engineering product development. Adv Eng Inform 52:101635. https://doi.org/10.1016/j.aei.2022.101635
4. Sharma A, Kosasih E, Zhang J, Brintrup A, Calinescu A (2022) Digital twins: state of the art theory and practice, challenges, and open research questions. J Ind Inf Integr 30:100383. https://doi.org/10.1016/j.jii.2022.100383
5. Lo CK, Chen CH, Zhong RY (2021) A review of digital twin in product design and development. Adv Eng Inform 48:101297. https://doi.org/10.1016/j.aei.2021.101297
6. Trauer J, Schweigert-Recksiek S, Engel C, Spreitzer K, Zimmermann M (2020) What is a digital twin?—Definitions and insights from an industrial case study in technical product development. Int Des Conf—Des. https://doi.org/10.1017/dsd.2020.15
7. Jiang Z, Guo Y, Wang Z (2021) Digital twin to improve the virtual-real integration of industrial IoT. J Ind Inf Integr 22:100196. https://doi.org/10.1016/j.jii.2020.100196
8. Kies AD, Krauß J, Schmetz A, Schmitt RH, Brecher C (2022) Interaction of digital twins in a sustainable battery cell production. Procedia CIRP 107:1216–1220. https://doi.org/10.1016/j.procir.2022.05.134
9. Stjepandic J, Sommer M, Denkena B (2022) DigiTwin: an approach for production process optimization in a built environment. Springer International Publishing, Switzerland. https://doi.org/10.1007/978-3-030-77539-1

10. Saadallah A, Finkeldey F, Buß J, Morik K, Wiederkehr P, Rhode W (2022) Simulation and sensor data fusion for machine learning application. Adv Eng Inform 52:101600. https://doi.org/10.1016/j.aei.2022.101600
11. Zhang X, Ming X, Bao Y, Liao X (2022) Industrial Internet Platform (IIP) enabled Smart Product Lifecycle-Service System (SPLSS) for manufacturing model transformation: from an industrial practice survey. Adv Eng Inform 52:101633. https://doi.org/10.1016/j.aei.2022.101633
12. Lind A, Högberg D, Syberfeldt A, Hanson L, Lämkull D (2022) Evaluating a digital twin concept for an automatic up-to-date factory layout setup. Adv Transdisciplinary Eng 21:473–484. https://doi.org/10.3233/ATDE220166
13. Ruppert T, Abonyi J (2020) Integration of real-time locating systems into digital twins. J Ind Inf Integr 20:100174. https://doi.org/10.1016/j.jii.2020.100174
14. Grau M, Korol W, Lützenberger J, Stjepandić J (2021) Automated generation of a digital twin of a process plant by using 3D scan and artificial intelligence. Adv Transdisciplinary Eng 16:93–102. https://doi.org/10.3233/ATDE210087
15. Sommer M, Stjepandić J, Stobrawa S (2021) Incremental update of a digital twin of a production system by using scan and object recognition. Adv Transdisciplinary Eng 16:83–92. https://doi.org/10.3233/ATDE210086
16. Guzina L, Ferko E, Bucaioni A (2022) Investigating digital twin: a systematic mapping study. Adv Transdisciplinary Eng 21:449–460. https://doi.org/10.3233/ATDE220164
17. Traoré MK (2021) Unifying digital twin framework: simulation-based proof-of-concept. IFAC-PapersOnLine 54(1):886–893. https://doi.org/10.1016/j.ifacol.2021.08.105
18. Grieves M, Vickers J (2017) Digital twin: mitigating unpredictable, undesirable emergent behavior in complex systems. In: Kahlen FJ, Flumerfelt S, Alves A (eds) Transdisciplinary perspectives on complex systems: new findings and approaches. Springer International, Cham, pp 85–113. https://doi.org/10.1007/978-3-319-38756-7_4
19. Shafto M et al (2010) Modeling, simulation, information technology and processing, DRAFT technology roadmap area Vol. 11. NASA, Washington. https://www.nasa.gov/pdf/501321main_TA11-MSITP-DRAFT-Nov2010-A1.pdf
20. Bu L, Zhang Y, Liu H et al (2021) An IIoT-driven and AI-enabled framework for smart manufacturing system based on three-terminal collaborative platform. Adv Eng Inform 50:101370. https://doi.org/10.1016/j.aei.2021.101370
21. Wang L, Liu Z, Liu A, Tao F (2021) Artificial intelligence in product lifecycle management. Int J Adv Manuf Technol 114:771–796. https://doi.org/10.1007/s00170-021-06882-1
22. Liu S, Lu Y, Zheng P, Shen H, Bao J (2022) Adaptive reconstruction of digital twins for machining systems: a transfer learning approach. Robot Comput-Integr Manufact 78:102390. https://doi.org/10.1016/j.rcim.2022.102390
23. Huang Z, Shen Y, Li J, Fey M, Brecher C (2021) AI-driven digital twins in industry 4.0: smart manufacturing and advanced robotics. Sensors 21:6340. https://doi.org/10.3390/s21196340
24. He B, Bai KJ (2021) Digital twin-based sustainable intelligent manufacturing: a review. Adv Manuf 9:1–21. https://doi.org/10.1007/s40436-020-00302-5
25. Li L, Lei B, Mao C (2022) Digital twin in smart manufacturing. J Ind Inf Integr 26:100289. https://doi.org/10.1016/j.jii.2021.100289
26. Zhang X, Hu B, Xiong G, Liu X, Dong X, Li D (2021) Research and practice of lightweight digital twin speeding up the implementation of flexible manufacturing systems. In: IEEE 1st international conference on digital twins and parallel intelligence (DTPI). https://doi.org/10.1109/DTPI52967.2021.9540104
27. Perno M, Hvam L (2020) Developing a framework for scoping digital twins in the process manufacturing industry. Adv Transdisciplinary Eng 13:475–486. https://doi.org/10.3233/ATDE200185
28. Stark R, Damerau T (2019) Digital twin. In: the international academy for production engineering. In: Chatti S, Laperrière L, Reinhart G, Tolio T (Hrsg) CIRP encyclopedia of production engineering, 2nd edn. Springer, Berlin Heidelberg. https://doi.org/10.1007/978-3-642-35950-7_16870-1

29. He B, Cao X, Hua Y (2021) Data fusion-based sustainable digital twin system of intelligent detection robotics. J Cleaner Prod 280:124181. https://doi.org/10.1016/j.jclepro.2020.124181
30. Wu M, Kozanoglu DC, Min C, Zhang Y (2021) Unraveling the capabilities that enable digital transformation: a data-driven methodology and the case of artificial intelligence. Adv Eng Inform 50:101368. https://doi.org/10.1016/j.aei.2021.101368
31. Bergs T, Gierlings S, Auerbach T, Klink A, Schraknepper D, Augspurger T (2021) The concept of digital twin and digital shadow in manufacturing. Procedia CIRP 101:81–84. https://doi.org/10.1016/j.procir.2021.02.010
32. Qi Q, Tao F, Hu T, Anwer N, Liu A, Wei Y, Wang L, Neef AYC (2021) Enabling technologies and tools for digital twin. J Manufact Syst 58(Part B):3–21. https://doi.org/10.1016/j.jmsy.2019.10.001
33. Ladj A, Wang Z, Meski O, Belkadi F, Ritou M, Da Cunha C (2020) A knowledge-based digital shadow for machining industry in a digital twin perspective. J Manuf Syst. https://doi.org/10.1016/j.jmsy.2020.07.018
34. Biahmou A, Emmer C, Pfouga A, Stjepandić J (2016) Digital master as an enabler for industry 4.0. Adv Transdisciplinary Eng 4:672–681. https://doi.org/10.3233/978-1-61499-703-0-672
35. Fukuda S, Lulić Z, Stjepandić J (2013) FDMU-functional spatial experience beyond DMU? In: Proceedings of the 20th ISPE international conference on concurrent engineering, CE 2013. IOS Press, Amsterdam, pp 431–440. https://doi.org/10.3233/978-1-61499-302-5-431
36. Goodwin T, Xu J, Celik N, Chen C-H (2022) Real-time digital twin-based optimization with predictive simulation learning. J Simul. https://doi.org/10.1080/17477778.2022.2046520
37. Kritzinger W, Karner M, Traar G, Henjes J, Sihn W (2018) Digital twin in manufacturing: a categorical literature review and classification. IFAC-PapersOnLine 51(11):1016–1022. https://doi.org/10.1016/j.ifacol.2018.08.474
38. Riesener M, Schuh G, Dölle C, Tönnes C (2019) The digital shadow as enabler for data analytics in product life cycle management. Procedia CIRP 80:729–734. https://doi.org/10.1016/j.procir.2019.01.083
39. Adamenko D, Kunnen S, Pluhnau R, Loibl A, Nagarajah A (2020) Review and comparison of the methods of designing the digital twin. Procedia CIRP 91:27–32. https://doi.org/10.1016/j.procir.2020.02.146
40. van der Valk H, Haße H, Möller F, Arbter M, Henning JL, Otto B (2020) A taxonomy of digital twins. In: Anderson B, Thatcher J, Meservy R (Hrsg.) Proceedings of the 26th Americas conference on information systems, pp 1–10
41. Semeraro C, Lezoche M, Panetto H, Dassisti M (2021) Digital twin paradigm: a systematic literature review. Comput Ind 130:103469. https://doi.org/10.1016/j.compind.2021.103469
42. Jones D, Snider C, Nassehi A, Yon J, Hicks B (2020) Characterising the digital twin: a systematic literature review. CIRP J Manuf Sci Technol 29:36–52. https://doi.org/10.1016/j.cirpj.2020.02.002
43. Leng J, Wang D, Shen W, Li X, Liu Q, Chen X (2021) Digital twins-based smart manufacturing system design in industry 4.0: a review. J Manuf Syst 60:119–137. https://doi.org/10.1016/j.jmsy.2021.05.011
44. Dittmann S, Zhang P, Glodde A, Dietrich F (2021) Towards a scalable implementation of digital twins—a generic method to acquire shopfloor data. Procedia CIRP 96:157–162. https://doi.org/10.1016/j.procir.2021.01.069
45. Sanchez-Londono D, Barbieri G, Fumagalli L (2022) Smart retrofitting in maintenance: a systematic literature review. J Intell Manuf. https://doi.org/10.1007/s10845-022-02002-2
46. Karnik N, Bora U, Bhadri K, Kadambi P, Dhatrak P (2022) A comprehensive study on current and future trends towards the characteristics and enablers of industry 4.0. J Ind Inf Integr 27:100294. https://doi.org/10.1016/j.jii.2021.100294
47. Conde J, Munoz-Arcentales A, Alonso A, López-Pernas S, Salvachúa J (2021) Modeling digital twin data and architecture: a building guide with FIWARE as enabling technology. IEEE Internet Comput 26(3):7–14. https://doi.org/10.1109/MIC.2021.3056923
48. Leng J, Liu Q, Ye S, Jing J, Wang Y, Zhang C, Zhang D, Chen X (2020) Digital twin-driven rapid reconfiguration of the automated manufacturing system via an open architecture model. Robot Comput-Integr Manufact 63:101895. https://doi.org/10.1016/j.rcim.2019.101895

49. Park KT, Lee J, Kim HJ, Noh SD (2020) Digital twin-based cyber physical production system architectural framework for personalized production. Int J Adv Manuf Technol 106:1787–1810. https://doi.org/10.1007/s00170-019-04653-7
50. DIN SPEC 91345: Referenzarchitekturmodell Industrie 4.0 (RAMI4.0). Deutsches Institut für Normung (DIN) e.V., 2016. https://doi.org/10.31030/2436156
51. Segovia M, Garcia-Alfaro J (2022) Design, modeling and implementation of digital twins. Sensors 22:5396. https://doi.org/10.3390/s22145396
52. Pan YH, Qu T, Wu NQ, Khalgui M, Huang GQ (2021) Digital twin based real-time production logistics synchronization system in a multi-level computing architecture. J Manufact Syst 58(Part B):246–260. https://doi.org/10.1016/j.jmsy.2020.10.015
53. Boyes H, Watson T (2022) Digital twins: an analysis framework and open issues. Comput Ind 143:103763. https://doi.org/10.1016/j.compind.2022.103763
54. Trauer J, Schweigert-Recksiek S, Schenk T, Baudisch T, Mörtl M, Zimmermann M (2022) A digital twin trust framework for industrial application. Proc Des Soc 2:293–302. https://doi.org/10.1017/pds.2022.31
55. Eickhoff T, Forte S, Göbel JC (2022) Approach for developing digital twins of smart products based on linked lifecycle information. Proc Des Soc 2:1559–1568. https://doi.org/10.1017/pds.2022.158
56. Moya B, Badías A, Alfaro I, Chinesta F, Cueto E (2022) Digital twins that learn and correct themselves. Int J Numer Meth Eng 123(13):3034–3044. https://doi.org/10.1002/nme.6535
57. Liu YK, Ong SK, Nee AYC (2022) State-of-the-art survey on digital twin implementations. Adv Manuf 10:1–23. https://doi.org/10.1007/s40436-021-00375-w
58. Leng J, Chen Z, Sha W, Lin Z, Lin J, Liu Q (2022) Digital twins-based flexible operating of open architecture production line for individualized manufacturing. Adv Eng Inform 53:101676. https://doi.org/10.1016/j.aei.2022.101676
59. Zhang Z, Guan Z, Gong Y, Luo D, Yue L (2022) Improved multi-fidelity simulation-based optimisation: application in a digital twin shop floor. Int J Prod Res 60(3):1016–1035. https://doi.org/10.1080/00207543.2020.1849846
60. Lucci N, Monguzzi A, Zanchettin AM, Rocco P (2022) Workflow modelling for human–robot collaborative assembly operations. Robot Comput-Integr Manufact 78:102384. https://doi.org/10.1016/j.rcim.2022.102384
61. Huang ZF, Soh KY, Islam MR, Chua KJ (2022) Digital twin driven life-cycle operation optimization for combined cooling heating and power-cold energy recovery (CCHP-CER) system. Appl Energy 324:119774. https://doi.org/10.1016/j.apenergy.2022.119774
62. Radanliev P, De Roure D, Nicolescu R, Huth M, Santos R (2022) Digital twins: artificial intelligence and the IoT cyber-physical systems in industry 4.0. Int J Intell Robot Appl 6:171–185. https://doi.org/10.1007/s41315-021-00180-5
63. Franceschi P, Mutti S, Ottogalli K, Rosquete D, Borro D, Pedrocchi N (2022) A framework for cyber-physical production system management and digital twin feedback monitoring for fast failure recovery. Int J Comput Integr Manuf 35(6):619–632. https://doi.org/10.1080/0951192X.2021.1992666
64. Feng B, Jiang M, Song K, Zhou C, Yu F, Wie X (2022) Research on intelligent warehouse digital twin system based on digital twin. In: Proceedings SPIE 12287, international conference on cloud computing, performance computing, and deep learning (CCPCDL 2022), 1228712. https://doi.org/10.1117/12.2641102
65. Wang G, Zhang G, Guo X, Zhang Y (2021) Digital twin-driven service model and optimal allocation of manufacturing resources in shared manufacturing. J Manuf Syst 59:165–179. https://doi.org/10.1016/j.jmsy.2021.02.008
66. Melesse TY, Di Pasquale V, Riemma S (2021) Digital twin models in industrial operations: state-of-the-art and future research directions. IET Collab Intel Manufact 3(1):37–47. https://doi.org/10.1049/cim2.12010
67. Laukotka FN, Rennpferdt C, Krause D (2022) Digital twins and product-service systems: a synergy with challenges and opportunities. Proc Des Soc 2:1639–1648. https://doi.org/10.1017/pds.2022.166

68. Xiong M, Wang H (2022) Digital twin applications in aviation industry: a review. Int J Adv Manuf Technol 121:5677–5692. https://doi.org/10.1007/s00170-022-09717-9
69. Rath N, Mishra RK, Kushari A (2022) Aero engine health monitoring, diagnostics and prognostics for condition-based maintenance: an overview. Int J Turbo Jet-Engines 0020. https://doi.org/10.1515/tjeng-2022-0020
70. Yang W, Zheng Y, Li S (2021) Application status and prospect of digital twin for on-orbit spacecraft. IEEE Access 9:106489–106500. https://doi.org/10.1109/ACCESS.2021.3100683
71. Franco da Silva AC, Wagner S, Lazebnik E, Traitel E (2022) Using a cyber digital twin for continuous automotive security requirements verification. IEEE Softw. https://doi.org/10.1109/MS.2022.3171305
72. Hu Z, Lou S, Xing Y, Wang X, Cao D, Lv C (2022) Review and perspectives on driver digital twin and its enabling technologies for intelligent vehicles. IEEE Trans Intell Veh 7(3):417–440. https://doi.org/10.1109/TIV.2022.3195635
73. Raza M, Prokopova H, Huseynzade S, Azimi S, Lafond S (2022) Towards integrated digital-twins: an application framework for autonomous maritime surface vessel development. J Mar Sci Eng 10(10):1469. https://doi.org/10.3390/jmse10101469
74. Liu L, Zhang X, Wan X, Zhou S, Gao Z (2022) Digital twin-driven surface roughness prediction and process parameter adaptive optimization. Adv Eng Inform 51:101470. https://doi.org/10.1016/j.aei.2021.101470
75. Zhu Q, Huang S, Wang G, Moghaddam SK, Lu Y, Yan Y (2022) Dynamic reconfiguration optimization of intelligent manufacturing system with human-robot collaboration based on digital twin. J Manuf Syst 65:330–338. https://doi.org/10.1016/j.jmsy.2022.09.021
76. Adu-Kankam KO, Camarinha-Matos LM (2022) A framework for the integration of IoT components into the household digital twins for energy communities. In: Camarinha-Matos LM, Ribeiro L, Strous L (eds) Internet of Things. IoT through a multi-disciplinary perspective. IFIPIoT 2022. IFIP advances in information and communication technology, vol 665. Springer, Cham. https://doi.org/10.1007/978-3-031-18872-5_12
77. Pratticò FG, CalandraD, Piviotti M, Lamberti F (2021) Assessing the user experience of consumer haptic devices for simulation-based virtual reality. In: IEEE 11th international conference on consumer electronics (ICCE-Berlin). https://doi.org/10.1109/ICCE-Berlin53567.2021.9719998
78. Kobayashi T, Fukae K, Imai T, Arai K (2022) Digital twin agent for super-aged society. In: IEEE international conference on consumer electronics (ICCE). https://doi.org/10.1109/ICCE53296.2022.9730230
79. Kurakova NG, Tsvetkova LA, Polyakova YV (2022) Digital twins in surgery: achievements and limitations. Khirurgiia 5:97–110. https://doi.org/10.17116/hirurgia202205197. PMID: 35593634
80. Flamini A, Loggia R, Massaccesi A, Moscatiello C, Martirano L (2022) BIM and SCADA integration: the dynamic digital twin. In: IEEE/IAS 58th industrial and commercial power systems technical conference (I&CPS). https://doi.org/10.1109/ICPS54075.2022.9773903
81. Kang JS, Chung K, Hong EJ (2021) Multimedia knowledge-based bridge health monitoring using digital twin (2021). Multimed Tools Appl 80:34609–34624. https://doi.org/10.1007/s11042-021-10649-x
82. Solman H, Kirch Kirkegaard J, Smits M, Van Vliet B, Bush S (2022) Digital twinning as an act of governance in the wind energy sector. Environ Sci Policy 127:272–279. https://doi.org/10.1016/j.envsci.2021.10.027
83. Turan E, Konuşkan Y, Yıldırım N, Tunçalp D, İnan M, Yasin O, Turan B, Kerimoğlu V (2022) Digital twin modelling for optimizing the material consumption: a case study on sustainability improvement of thermoforming process. Sustain Comput: Inform Syst 35:100655. https://doi.org/10.1016/j.suscom.2022.100655
84. Conde J, Munoz-Arcentales A, Romero M, Rojo J, Salvachua J, Huecas G, Alonso Á (2022) Applying digital twins for the management of information in turnaround event operations in commercial airports. Adv Eng Inform 54:101723. https://doi.org/10.1016/j.aei.2022.101723

85. Xia H, Liu Z, Efremochkina M, Liu X, Lin C (2022) Study on city digital twin technologies for sustainable smart city design: a review and bibliometric analysis of geographic information system and building information modeling integration. Sustain Cities Soc 84:104009. https://doi.org/10.1016/j.scs.2022.104009
86. Bamberg A, Urbas L, Bröcker S, Bortz M, Kockmann N (2021) The digital twin—your ingenious companion for process engineering and smart production. Chem Eng Technol 44:954–961. https://doi.org/10.1002/ceat.202000562
87. Shcherbatov I, Agibalov V, Dolgsuhev A, Belov M (2022) Subsystem for building a digital twin of the main and auxiliary equipment of thermal scheme of thermal power plant. In: Kravets AG, Bolshakov AA, Shcherbakov M (eds) Society 5.0: human-centered society challenges and solutions. studies in systems, decision and control, vol 416. Springer, Cham. https://doi.org/10.1007/978-3-030-95112-2_20
88. Udugama IA, Lopez PC, Gargalo CL, Li X, Bayer C, Gernaey K (2021) Digital twin in biomanufacturing: challenges and opportunities towards its implementation. Syst Microbiol Biomanuf 1:257–274. https://doi.org/10.1007/s43393-021-00024-0
89. Martínez GS, Sierla SA, Karhela TA, Lappalainen J, Vyatkin V (2018) Automatic generation of a high-fidelity dynamic thermal-hydraulic process simulation model from a 3D plant model. IEEE Access 6:45217–45232. https://doi.org/10.1109/ACCESS.2018.2865206
90. Xavier M, Håkansson J, Patil S, Vyatkin V (2021) Plant model generator from digital twin for purpose of formal verification. In: 26th IEEE international conference on emerging technologies and factory automation (ETFA). pp 1–4. https://doi.org/10.1109/ETFA45728.2021.9613704
91. Maheshwari P, Kamble S, Belhadi A, Mani V, Pundir A (2022) Digital twin implementation for performance improvement in process industries—a case study of food processing company. Int J Prod Res. https://doi.org/10.1080/00207543.2022.2104181
92. Huynh TA, Zondervan E (2022) Process intensification and digital twin—the potential for the energy transition in process industries. Phys Sci Rev. https://doi.org/10.1515/psr-2022-0058
93. Hu S, Wang S, Su N, Li X, Zhang Q (2021) Digital twin based reference architecture for petrochemical monitoring and fault diagnosis. Oil Gas Sci Technol—Rev IFP Energies nouvelles 76:9. https://doi.org/10.2516/ogst/2020095
94. Azangoo M, Sorsamaki L, Sierla SA, Matasniemi T, Rantala M, Rainio K, Vyatkin V (2022) A methodology for generating a digital twin for process industry: a case study of a fiber processing pilot plant. IEEE Access 10:58787–58810. https://doi.org/10.1109/ACCESS.2022.3178424
95. Eswaran M, Raju Bahubalendruni MVA (2022) Challenges and opportunities on AR/VR technologies for manufacturing systems in the context of industry 4.0: a state of the art review. J Manuf Syst 65:260–278. https://doi.org/10.1016/j.jmsy.2022.09.016
96. Bevilacqua M, Bottani E, Ciarapica FE, Costantino F, Di Donato L, Ferraro A, Mazzuto G, Monteriù A, Nardini, G, Ortenzi M, Paroncini M, Pirozzi M, Prist M, Quatrini E, Tronci M, Vignali G (2020) Digital twin reference model development to prevent operators' risk in process plants. Sustainability 12:1088. https://doi.org/10.3390/su12031088
97. Pfouga A, Stjepandić J (2018) Leveraging 3D geometric knowledge in the product lifecycle based on industrial standards. J Comput Des Eng 5(1):54–67. https://doi.org/10.1016/j.jcde.2017.11.002
98. Czerniawski T, Leite F (2020) Automated digital modeling of existing buildings: a review of visual object recognition methods. Autom Constr 113:103131. https://doi.org/10.1016/j.autcon.2020.103131
99. Mirzaei K, Arashpou M, Asadi E, Masoumi H, Bai Y, Behnood A (2022) 3D point cloud data processing with machine learning for construction and infrastructure applications: a comprehensive review. Adv Eng Inform 51:101501. https://doi.org/10.1016/j.aei.2021.101501
100. Kawashima K, Karnai S, Date H (2013) As-built modeling of piping system from terrestrial laser-scanned point clouds using normal-based region growing. J Comput Des Eng 1:13–26. https://doi.org/10.7315/JCDE.2014.002

101. Maalek R, Lichti DD, Walker R, Bhavnani A, Ruwanpura JY (2019) Extraction of pipes and flanges from point clouds for automated verification of pre-fabricated modules in oil and gas refinery projects. Autom Constr 103:150–167. https://doi.org/10.1016/j.autcon.2019.03.013
102. Cheng L, Wei Z, Sun M, Xin S, Sharf A, Li Y, Chen B, Tu C (2020) DeepPipes: learning 3D pipelines reconstruction from point clouds. Graph Models 111:101079. https://doi.org/10.1016/j.gmod.2020.101079
103. Agapaki E, Brilakis I (2020) CLOI-NET: class segmentation of industrial facilities' point cloud datasets. Adv Eng Inform 45:101121. https://doi.org/10.1016/j.aei.2020.101121
104. Lee ID, Lee I, Han S (2021) 3D reconstruction of as-built model of plant piping system from point clouds and port information. J Comput Des Eng 8(1):195–209. https://doi.org/10.1093/jcde/qwaa072
105. Braun D, Schloegl W, Weyrich M (2021) Automated data-driven creation of the digital twin of a brownfield plant. In: 26th IEEE international conference on emerging technologies and factory automation (ETFA). pp 1–7. https://doi.org/10.1109/ETFA45728.2021.9613676
106. Rantala M, Niemistö H, Karhela T, Sierla S, Vyatkin V (2019) Applying graph matching techniques to enhance reuse of plant design information. Comput Ind 107(2019):81–98. https://doi.org/10.1016/j.compind.2019.01.005
107. Sierla S, Azangoo M, Rainio K, Papakonstantinou N, Fay A, Honkamaa P, Vyatkin V (2022) Roadmap to semi-automatic generation of digital twins for brownfield process plants. J Ind Inf Integr 27:100282. https://doi.org/10.1016/j.jii.2021.100282
108. Sierla S, Pourakbari-Kasmaei M, Vyatkin V (2022) A taxonomy of machine learning applications for virtual power plants and home/building energy management systems. Autom Constru 136:104174. https://doi.org/10.1016/j.autcon.2022.104174
109. Schweigert-Recksiek S, Trauer J, Engel C, Spreitzer K, Zimmermann M (2020) Conception of a digital twin in mechanical engineering—a case study in technical product development. Int Des Conf—Des 383–392. https://doi.org/10.1017/dsd.2020.23
110. Kong T, Hu T, Zhou T, Ye Y (2021) Data construction method for the applications of workshop digital twin system. J Manuf Syst 58:323–328. https://doi.org/10.1016/j.jmsy.2020.02.003
111. Fang X, Wang H, Liu G, Tian X, Ding G, Zhang H (2022) Industry application of digital twin: from concept to implementation. Int J Adv Manuf Technol 121:4289–4312. https://doi.org/10.1007/s00170-022-09632-z
112. Gürdür Broo D, Bravo-Haro M, Schooling J (2022) Design and implementation of a smart infrastructure digital twin. Autom Constr 136:104171. https://doi.org/10.1016/j.autcon.2022.104171
113. Gao L, Jia M, Liu D (2022) Process digital twin and its application in petrochemical industry. J Softw Eng Appl 15:308–324. https://doi.org/10.4236/jsea.2022.158018
114. Zhang D, Gao X (2022) A digital twin dosing system for iron reverse flotation. J Manuf Syst 63:238–249. https://doi.org/10.1016/j.jmsy.2022.03.006
115. Eppinger T, Longwell G, Mas P, Goodheart K, Badiali U, Aglave R (2021) Increase food production efficiency using the executable digital twin (XDT). Chem Eng Trans 87:37–42. https://doi.org/10.3303/CET2187007
116. Lázaro O et al (2022) Model-based engineering and semantic interoperability for trusted digital twins big data connection across the product lifecycle. In: Curry E, Auer S, Berre AJ, Metzger A, Perez MS, Zillner S (eds) Technologies and applications for big data value. Springer International Publishing, Cham. https://doi.org/10.1007/978-3-030-78307-5_18
117. Birk W, Hostettler R, Razi M, Atta K, Tammia R (2022) Automatic generation and updating of process industrial digital twins for estimation and control—a review. Front Control Eng 3. https://doi.org/10.3389/fcteg.2022.954858
118. Kremer P, Lützenberger J, Müller F, Stjepandić J (2022) An approach for the incremental update of a digital twin of a process plant. Adv Transdisciplinary Eng 28:310–319. https://doi.org/10.3233/ATDE220660

Chapter 4
Common Practice in Plant Design with Interoperability Standards

Abstract Plant design as a phase of plant engineering comprises the activities in the conception and the planning of plants where chemical substances, biological agents, or auxiliary energy from available raw materials or existing energy sources are typically produced. Plant construction projects are usually defined as so-called engineering, procurement, and construction (EPC) projects which include the provision of the engineering design, the procurement of the necessary materials needed for construction, and finally the actual construction. EPC contractors are responsible for planning, designing, manufacturing/assembling, and handover the plant to the operator. Hereby, planning activities of a plant are conducted by using CAD systems for mechanical engineering, civil engineering, and electrical engineering as well as ERP and PDM systems. In this context, the application scenario "Seamless and dynamic engineering of plants" (SDP) is explained and the overall principle is presented as an initial engineering process for engineering and construction of a plant. Here, an integrating plant model is created, which is maintained and kept consistent throughout the entire life of the real physical plant in permanently interrelated processes between engineering, operation, and service of the plant. Ever deeper standardization in Industry 4.0 entails tool vendor-neutral representations of piping and instrumentation diagrams (P&ID) as well as 3D pipe routing. For the sake of the digital twin, a complete digital plant model requires combining these two representations. 3D pipe routing information is essential for building any accurate first-principles process simulation model, in particular for a brownfield plant. Piping and instrumentation diagrams are explored as the primary source for control loops. The interoperability as a fundamental capability of the digital twin is the key constituent of almost every initiative and approach. An excerpt of the initiatives, the frameworks, and the standards regarding the digital twin in a process plant is highlighted. In the discussion section, the global goals of the digital twin are presented, and how they can be achieved in the case of a brownfield plant. Finally, the perspective of working with the digital twin of a brownfield plant is drafted.

Keywords Plant engineering · Plant design · EPC contractor · Brownfield plant · Piping engineering · P&ID · Interoperability · CAD · PDM

J. Stjepandić et al., *Generation and Update of a Digital Twin in a Process Plant*,
https://doi.org/10.1007/978-3-031-47316-6_4

4.1 Introduction

The aim of plant engineering is the conception, planning, implementation, and operation of industrial plants by creating the technical prerequisites for conducting a process consisting of several steps in order to transform raw material and energy into products that are useful to society, at an industrial level [1]. Conversely, plant design comprises activities in the conception and planning of plants [2]. In such plants, chemical substances, biological agents, or auxiliary energy from available raw materials or existing energy sources are typically produced. Additionally, plant engineering also provides complex automated production systems to the industry.

A plant is understood as a system of coordinated technical equipment that forms a unit and emerges as an outcome of the planning and construction process. The formal foundation of a plant is the necessary permissions from public bodies which are based on the comprehensive documentation for the construction and assembly work. The documentation comprises the description of components, the assembly plans for buildings, equipment, piping, fittings, and cabling as well as the logic plans for the control and regulation of the processes, auxiliary equipment, and, finally, operating instructions. Piping and instrumentation diagrams (P&ID) are mandatory. In some countries, 3D models of pipeline systems are mandatory as well [3].

Plant construction projects are usually defined as so-called engineering, procurement, and construction (EPC) projects which include the provision of the engineering design, the procurement of the necessary materials needed for construction, and finally the actual construction. One of the major challenges in plant construction is coordination. A plant construction project is completed with commissioning, proof of the required performance parameters, and handover to the operator. These tasks are taken by a general contractor who is referred to as the EPC contractor. The client closes a contract with the EPC contractor, who then delivers a plant ready to service to the client [3].

In comparison with the traditional design-bid-build (DBB) method, the EPC facilitates a fast-track project delivery. The client stays concurrently involved and can shorten the timeframe to execute planning, procurement, and maintenance activities. EPC projects ensure that the EPC contractor plans, conducts, and finalizes the entire project, hiring the actors of the supply chain (the manufacturer, designer, contractor, and subcontractor) to create a project team to meet the EPC goals (Fig. 4.1). The responsibility of the EPC contractors includes the compliance with regulatory requirements and provision of all appropriate measures to protect the environment that is very important to governments' acceptance and local people's interests. Furthermore, the EPC contractor needs to cooperate with all involved social-political parties (e.g., central government, local authorities, and residents/communities) to get resources, approval, and support to approach the project's success [3].

Figure 4.1 depicts the share of work, roles, and responsibilities in the plant industry. The EPC contractor is responsible for planning, design, manufacturing/ assembling, and handover of the plant to the operator. The planning activities are conducted by using CAD systems for mechanical engineering, civil engineering, and

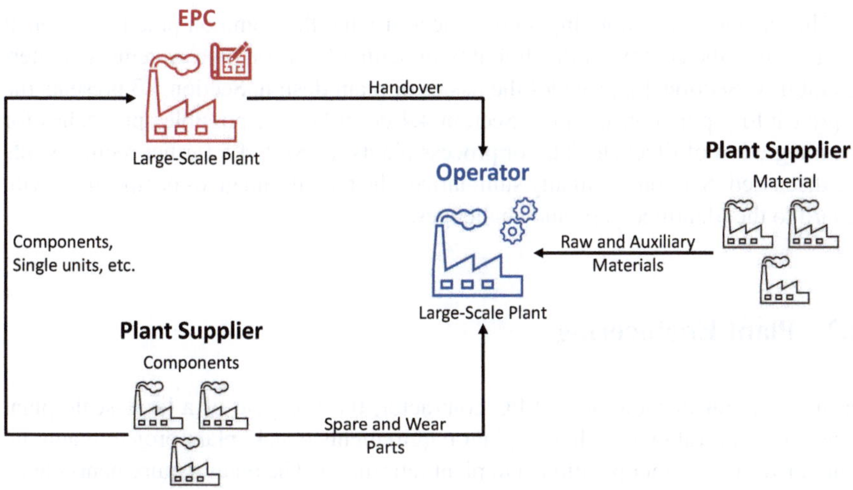

Fig. 4.1 EPC and owner in the plant industry

electrical engineering as well as ERP and PDM systems. The operator is usually the owner of the plant and is responsible for the operations and maintenance of the plant by using systems like SCADA, and MES which provide and prepare process data for process optimization. The plant supplier is responsible for the design, development, manufacturing, and delivery of components and sub-systems to the EPC and also for supplying the operator with spare and wear parts. Such suppliers work with CAD systems for mechanical engineering as well as with ERP and PDM. Further suppliers are responsible for supply with raw and auxiliary materials. During the handover, the EPC contractor delivers documentation, manuals, and service plans to the operator [4].

Process Systems Engineering (PSE) aims at the integration of scales and components describing the behavior of a physicochemical system, via mathematical modeling, data analytics, design, optimization, and control. PSE offers a scientific basis and computational tools for addressing contemporary and future challenges for process engineering problems from different domains such as energy, environment, the ‘industry of tomorrow’, and sustainability. PSE encompasses a three-layered view: (1) the inner core fundamental layer (process–product related activities where the application of the fundamental concepts of PSE help to design, build and operate manufacturing processes that convert specific raw materials to desired products), (2) the expanding layer (resources-efficiency related activities where improved understanding of the concepts and combination of science and engineering lead to the development of new technologies that when applied, lead to more sustainable engineering solutions), and (3) the outer unifying layer (activities that impact societal challenges where industrial development helps to address challenges approaching a more sustainable society) [5].

This chapter aims at giving an overview of what the common practices in plant design are in the context of the digital twin, with a focus on pipe systems and interoperability. Section 4.2 provides the basics of plant design. Section 4.3 presents the approach to piping construction. Section 4.4 describes the possible approaches for the integration of lifecycle data for process plants. In Sect. 4.5 further requirements are discussed. Section 4.6 finally summarizes the results and gives perspectives with regard to the identified gaps and challenges.

4.2 Plant Engineering

From the point of view of an EPC contractor, the lifecycle of a large-scale plant consists of the following phases: plant requirements check, plant project planning, plant provision, plant operation, and plant retirement. The plant requirements check encompasses the project initiation, the analysis of market trends, market changes, and customer demands, and check for realizability (technically and economically), and potential collaborations. Plant project planning comprises prequalification, quotation preparation, basic engineering, documentation, cooperation building, negotiation, and contract design. Plant provision consists of project management, detailed design, project procurement, manufacturing, transport, assembly, commissioning, and documentation. Plant operation encompasses marketing support, plant operation, plant maintenance, and plant overhaul. Finally, plant retirement consists of recycling, reconditioning, and disposal.

The objective of basic engineering is the creation of a feasibility concept under consideration of manufacturing technology, process engineering, and energy technology. Several boundary conditions need to be considered: which output of the plant is expected; part quantities, substance quantities, and energy quantities; what is the entitlement of the output; quality parameters; what are conditions on-site; and the given infrastructure. In the scope of the basic engineering lie aspects such as all process steps, necessary technical equipment, supply and disposal equipment, necessary control and regulation technology, options for maintenance and servicing, safety techniques [6], necessary permits, and, finally, expected costs and revenues.

Detailed engineering focuses on the detailed design of the overall system considering aspects such as the procurement of all permits, concrete design and linking of the components, contract design with subcontractors, elaboration, and compilation of the documentation, preparation of flow charts, preparation of operation instructions (focus: commissioning, decommissioning, behavior in case of malfunction), preparation of maintenance and servicing plans, and creation of an operating concept.

4.2.1 Process Plant Engineering

The process in a plant consists of the main process and ancillary processes (Fig. 4.2). Within the main process raw materials are transformed into final products by using auxiliary materials and energy. The following steps need to be considered: design of stocks for raw and auxiliary materials depending on the substance type (pressure vessels, tank facilities, bulk storage, etc.), the quantity of substances, operating conditions (pressure, temperature, etc.), substance properties, the preparation of raw and auxiliary materials (shredding, mixing, dosing) and feeding them into reaction stages of the main process. Further steps are the separation of intermediate products between the reaction stages from the responsive, storage of intermediate products, consideration of influencing variables (substances, temperature, pressure) of reaction stages, definition of limit values for influencing variables, separation of the final product and the reaction partners, consideration of regulation and control technology, conveyer technology, dosing technology, etc. [1].

The objective of the ancillary processes is the safeguarding and maintaining of the main process. Necessary ancillary processes are waste treatment, energy supply, supply with raw and auxiliary materials, delivery of raw and auxiliary materials, delivery of the final product, and connection with the infrastructure.

The layout of machines and apparatuses needs to fulfill the following requirements:

- Guarantee of continuous operation,
- Failure probabilities of the individual components must be approximately equal,
- The subsystem with the highest probability of failure determines the maintenance cycle,
- Redundancies need to be considered if necessary (e.g., installation of reserve pumps connected in parallel).

The most important technologies used for the main process and ancillary processes in a process plant are [7]:

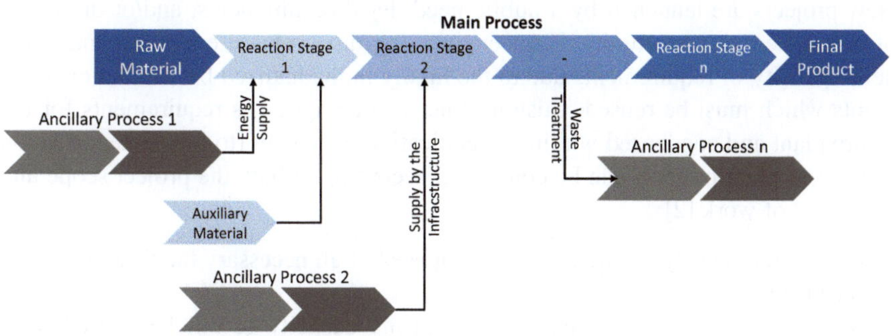

Fig. 4.2 Process of the process plant engineering

- Stock technology (tank farms, pressure vessels, bulk storage; above and below ground)
- Conveyer and dosing technology (piping systems, pumps, compressor, screw conveyors, conveyor belts)
- Fitting technology (shut-off valves, non-return-valves, flow control valves, actuator valves)
- Mixing technology (stirring machines, jet pumps, close-clearance mixers, screw blenders)
- Reactor technology (Single-phase reactors, multiphase reactors, stirred tanks, tubular reactors)
- Heat generation technology (steam boiler, hot water boiler, natural circulation boiler)
- Heat transfer technology (heating/cooling coils, heating/cooling jackets, cooling towers, regenerators)
- Substance separation technology (centrifuges, separators, columns, filters, cyclones, sedimentation tanks)
- Automation technology (regulation and control units, process control monitoring)
- Ventilation and air conditioning systems (fans, filtration, cleaning, cooler).

While most technologies used in a process plant work with fluids and bulk goods, which need a mutual interconnection, the piping systems of any complexity are an indispensable constituent of each unit in a plant and play an important role in the definition of the layout of a plant due to its variability. Regarding the operation, maintenance, retrofit, and overhaul, the piping systems gain particular importance because almost any change in the equipment and position causes a change in the piping system which must be planned appropriately. That is why the approach presented here will be focused on the recognition of piping systems and their components.

4.2.2 Plant Project Planning

New projects are launched by a public need, legal requirements, and/or discovery of a commercial opportunity. Plant engineering discovers an enormous mass and heterogeneity of requirements due to the highly individualized brownfield environments which must be reused. Customer needs are aligned as requirements for the future plant and condensed within a specification document (to-be). Categorization of process plant projects can be conducted according to both the project scope and the extent of work [2]:

- Completely new (greenfield) plant composed of all necessary facilities and units on a new site
- Revision of an existing plant for increasing capacity or efficiency (debottlenecking)
- Modernization (e.g., upgrading equipment or instrumentation/automation)
- Overhaul by one or more new units on an existing plant site (brownfield projects).

A brief overview of sections of a built plant can help figure out what kind of scope an industrial plant project may potentially encompass. The sections that make up a process plant are most commonly the following [2]:

- Process unit(s)
- Raw material receipt, storage, and transport
- Product storage and dispatch
- Utilities/off-sites and interconnecting
- Automation and control systems
- Electrical systems
- Drain/sewer/filter/purification systems and waste treatment
- Safety, security, and protection systems
- Buildings, structures, and road/railway/port.

Process units refer to equipment units assembled to produce intermediate or final products which include supplementary systems (electrical equipment, structures, field instrumentation, cabling, pipeline systems, and material transport). Liquid and gas processing units most commonly consist of, as major equipment, heat exchangers, tanks, reactors, furnaces, pressure vessels, towers, pumps, compressors, turbines, and various packaged systems, with extensive onsite piping runs and connections having a variety of on-line appurtenances. Solid processing (solid–liquid or solid–gas) plants or such parts of process plants additionally contain some other types of major equipment, mostly for conveying and storing solids, like conveyors, elevators, bunkers, chutes, rotary feeders, cyclone separators, dust collection filters, crushers/grinders and kilns [2]. In the case of revision, modernization, and overhaul, one or more process units and related pipeline systems are the subjects of change.

In many areas of plant engineering, a process plant for a certain purpose is first pre-developed by the EPC contractor independently of the customer. This is the so-called "reference plant". Subsequently, the reference plant can be adjusted and integrated with different customers. However, most projects in plant construction are not greenfield but brownfield projects, where construction takes place in existing buildings, and existing infrastructure must be considered. This provides additional complexity for the project planning. The motivation is therefore to identify the customer's requirements as efficiently as possible and to process them automatically. For this purpose, a library with relevant geometric entities of the customer's brownfield environment can be beneficial [8].

4.2.3 Asset Overhaul

Plants are optimized regarding assets and resources. In an ideal case, when the demand for a single product can justify its own manufacturing facility that is operating at full capacity, a separate facility can be built for that product. This decision is made based on a runner/repeater/stranger (RRS) analysis if that product is considered a runner [9]. Otherwise, most products will likely start out as strangers, attempting to catch repeaters, and hopefully become runners eventually.

In common practice, vendors mostly install multi-product plants, which have the flexibility to run new processes in existing, different plants with low or no additional capital investment. Conversely, the process operation in existing plants can yield a suboptimal technical performance, which is different from a suboptimal performance from a business perspective [10]. Therefore, the multi-product plants are inherently candidates for overhaul.

Leveraging existing plants is not always the best option, especially when the resulting suboptimal process incurs additional, hardly predictable costs, e.g., quality or control means. This approach also sets the minimum cost for a given company to produce a dedicated product, while a company with different plant options may be able to produce the same product at a significantly lower internal cost. This can impose make-or-buy decisions on a particular product when an outside supplier can deliver more cheaply than using internal resources [10].

This is the starting point for an overhaul of one or more units in a plant. When an organization has the knowledge necessary for a process to understand the impact that changing some units or equipment will have on the overall manufacturing cost, the business has a number of options to gain a competitive advantage. This can be achieved by integrating new units into the existing available equipment or, if the benefits can be quantified, and the investment justified, by setting up an entirely new plant [10].

The decision whether to use existing or to build new plants is often driven by broader business considerations, such as significant downtimes, availability of fully depreciated plants, supply chain integration (e.g., just-in-time, production for storage), available resources (e.g., dealing with new technologies) and knowledge development/maintenance. This also needs to be compensated with the level of acceptable risk within an organization and this is often a subject of much debate. This is the field where an overhaul is being considered because the risk seems manageable. With the development of manufacturing technologies, there are often a variety of novel technologies that have the technical promise of reducing overall costs and improving quality. Successful organizations are those that manage their exposure to unproven technologies by identifying those that have the greatest potential impact (across multiple products in the organization's portfolio) and directing attention to those areas. Specialized technologies that have a narrower range of future applicability pose a higher risk than technologies with great future potential [10].

4.2.4 Seamless and Dynamic Engineering of Plants

As presented in the previous section, engineering work is not only performed in the initial design and construction of a plant but will be necessary throughout the whole lifetime of a plant, to adjust to the dynamics [11]. The digital twin is a significant supporting means to ensure a seamless process between involved organizations. Upon a request by a public body, the usage of a digital plant model throughout the whole lifetime of a plant is elaborated by a technical committee in a guideline [11] in a generic way to provide a better understanding in a broad community.

A major goal of Industry 4.0 is to make plant information available to humans and machines throughout the network of enterprises involved in designing, commissioning, and operating the plant [12]. For this purpose, the application scenario "Seamless and dynamic engineering of plants" (SDP) was developed. The overall principle of the application scenario SDP is that in an initial engineering process for the engineering and construction of a plant, an integrating plant model is created, which is maintained and kept consistent throughout the entire life of the real physical plant in permanently interrelated processes between engineering, operation, and service of the plant.

The concepts of usage of an integrating plant model can be applied to both greenfield and brownfield projects. Also, these concepts can be applied both in discrete industries and in process industries, even if it is characterized by terms that are used more in process industries than in discrete industries [11].

In all of these cases models, libraries, and physical objects typically already exist. In a usage view, an integrating plant model is described in terms of its application perspective. Although an integrating plant model contains structural, functional, and behavior-based aspects, these aspects are not considered in a usage view but should be considered in a functional view, which is beyond the scope of this book [11].

From a business view, we distinguish between value chains, which are based on an exchange of models (which is in the hands of operators who are integrating the plant model and, thus, important for our approach), and the value chain of the software supplier, which is based on the provision of a tool environment. Value chains between stakeholders based on the provision of physical assets are added to the business view [11].

The overall principle of the application scenario SDP (Fig. 4.3) is that in an initial engineering process for the engineering and construction of a plant, an integrating plant model is created, which is maintained and kept consistent throughout the entire life of the real physical plant in permanently interrelated processes between engineering, operation, and service of the plant. Besides a model of the real physical plant over its lifetime (engineering and operation phase including conversions), this model also includes boundary conditions, context information, possible variants of the plant, conceivable and implemented engineering decisions as well as the impact of such decisions [11].

For a successful implementation of an integrated plant model across the complete life of a plant, including all involved stakeholders, it is necessary to create a common understanding of the model and the benefits it provides, by a usage view [12], based on the usage viewpoint proposed by the Industrial Internet Consortium [13]. The tasks according to this architecture include a functional map referring to the functional viewpoint and an implementation map to the implementation viewpoint.

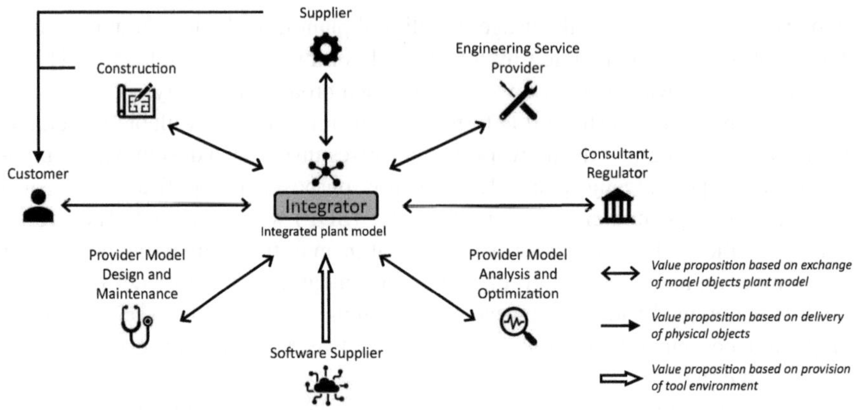

Fig. 4.3 Value network according to the business view of application scenario SDP [11]

Regarding the system under consideration, objects are assigned—because of their fundamental being—to either the physical or the model world. Regardless of this assignment, every object exists in reality and has its own life. However, the management of the objects and their life are fundamentally different. While physical objects have their own physical life, exist regardless of whether they are somehow considered or not, and are permanently in an aging process due to their physical presence, objects of the model world remain unchanged in themselves. They change only because of targeted interventions from outside [11].

Objects of the model world are pure information objects. However, these cannot exist virtually but need a physical carrier. But the information object itself, however, is completely independent of its physical carrier and does not participate in the life of the physical carrier. Conversely, objects of the model world are independent of the systems they might describe in the physical world [11].

The concept of model object forms the basis for modeling [11]:

- A model object can be associated with another model object via a relation. A specific relation between model objects is an aggregation, where the existence of the sub-models is independent of the existence of the superior model object.
- Model objects may have attributes and attributes have values. Attributes may have additional characteristics like value range, unit, or default value.
- The concept of a model object template is intended to be used for the creation of model objects. A model object template describes common characteristics of model objects, which are created from the model object template. The model object templates and their lives are considered as part of the system under consideration.

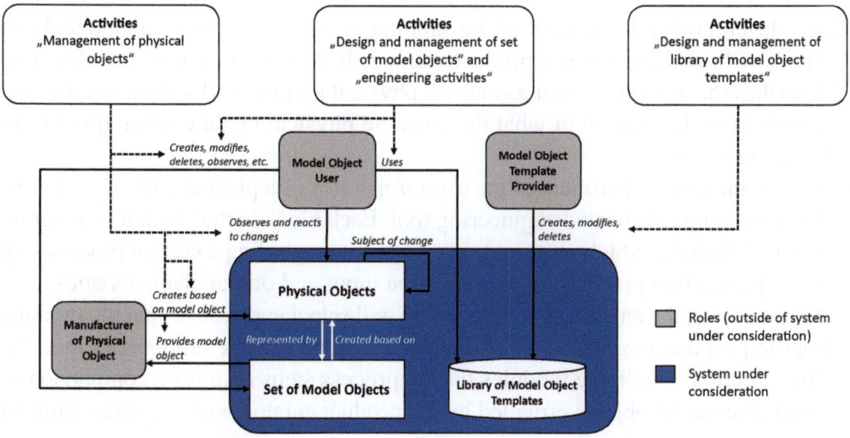

Fig. 4.4 Clusters of activities in connection with the system under consideration and roles [11]

The core activities as illustrated in Fig. 4.4 are broken down into the following different clusters: (a) Design and management of a set of model objects, (b) Design and management of a library of model object templates, (c) Management of physical objects, and (d) Engineering activities. A separate cluster can be added for the activities related to the management of a tool environment, where some activities with respect to the development and operation of a tool environment are drawn [11].

The concept of a model object template is dedicated to being used for the creation of model objects. A model object template describes common characteristics of model objects which are created from the model object template. The model object templates and their lives are considered part of the system under consideration. Model object templates can be structured by relations and can be organized in libraries. Each model object template and each library have an owner and a lifecycle [13].

As an example, the engineering of modular plants is currently discussed as the typical practice case in the context of module-type packages (MTP). There is the supplier of a module, who develops a model series of a module and, on request of a customer, produces and delivers a corresponding physical module to the customer. The supplier of the module provides an interface description of the module following the specification according to MTP. Such a module could be, for example, a reactor or a mixing unit. On the other hand, the owner/operator of a plant, who first designs a plant, in which such a module will be installed, then orders a corresponding module and physically installs this module in the plant. Because the supplied module complies with standardized interfaces, the installation and replacement of such a module in the infrastructure can be done easily. The infrastructure comprises both physical aspects such as material flow and energy supply, as well as the integration into the application software of the process control system [11].

The engineering of the module and the engineering of the plant are the following characteristics [11]:

- The development department of the supplier of a module creates a model object which will also comprise a product catalog. It is a model object template that describes the assurances that a concrete physical module will satisfy. In addition, it includes a description of what the concrete physical module will require from its environment.
- The engineering department of the owner/operator of a plant creates a block flow diagram using a dedicated engineering tool. Each block in the block flow diagram is a model object which represents requirements regarding a certain process step of the production process, for example, the input and output characteristics.
- At some time, the engineering department will select and order a specific module. For this purpose, the product catalog is made available by the supplier of a module. The selection of suitable modules for the process steps requires a comparison of respective model objects provided by the product catalog with the various model objects.

To facilitate the selection decision of an owner/operator of a plant, who wants to use the module, the supplier of a module provides a so-called "Module Type Package (MTP)", which is a model object containing a subset of the information of the model object (Fig. 4.5). It has been created and is made available to the engineering department of the owner/operator of a plant as module_MTP based on the selection of the owner/operator of a plant in the product catalog. The structure and general content of a MTP are described in a model object template MTP, as specified jointly by NAMUR (owner/operator organization) and ZVEI (supplier organization). The engineering department of the owner/operator of a plant will combine module_ MTP with the model objects representing the MTPs of the other process steps and check (by using a tool environment) for static and dynamic compatibility [11].

Once the engineering department of the owner/operator of a plant has decided to order a certain module according to module_MTP (and probably additional information) for the plant, a physical representation physical_module_MTP of the model

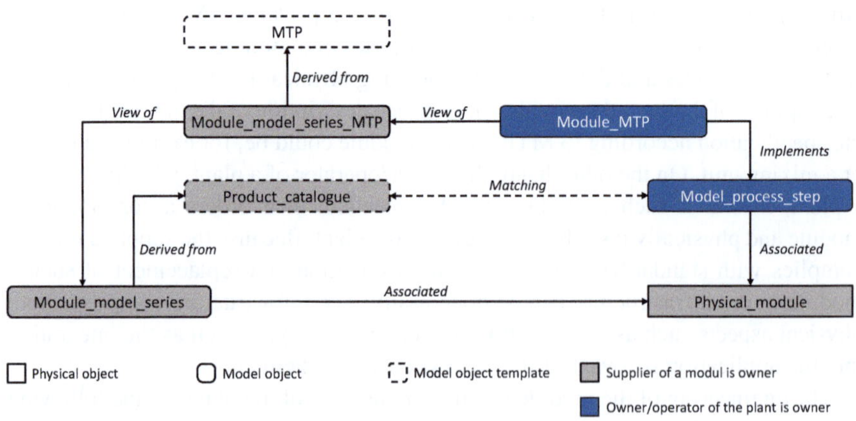

Fig. 4.5 Overview of main physical objects, model objects, and model object templates [11]

object module_MTP is loaded into the process control system (Fig. 4.5). The creation of the physical object physical_module_MTP is done and the result could be a physical file according to a certain file format, for example, XML Schema [11].

The process control system—as a specific exemplification of a tool environment with the main physical objects, model objects, model object templates, and relationships—recognizes this change in the physical world and will react correspondingly. For example, appropriate model objects are created so that on the basis of these model objects the process control system is prepared by generating physical software in the physical process control system that after the physical plug of the physical object into the physical process control system communication between the physical object and the physical process control system is possible [11].

The physical module delivered by the supplier of the module is physically integrated into the plant, for example, placed at a suitable place and connected via pipes and cables with the backbone of the plant. Thus, the physical objects are connected according to the possibilities for material, energy, and information flow.

4.3 Basics of Piping Construction

Ever deeper standardization in Industry 4.0 entails tool vendor-neutral representations of piping and instrumentation diagrams (P&ID) as well as 3D pipe routing. To build a digital twin, a complete digital plant model requires combining these two representations. 3D pipe routing information is essential for building any accurate first-principles process simulation model, in particular for a brownfield plant. Piping and instrumentation diagrams are the primary sources for control loops [14]. Flow diagrams are among the documents of crucial importance for a process plant. They are generated in three levels of detail:

1. Basic flow diagram
2. Process flow diagram (PFD)
3. Piping and instrumentation diagram (P&ID).

While the basic and process flow diagrams are created as part of the basic engineering, the detailed pipeline and instrument flow diagrams (P&IDs) are only processed after the order for the execution of the system has been issued as part of the detailed engineering. Regulations for the creation of flow diagrams are either based on the customer specifications or are agreed between the operator and the plant manufacturer [15].

In order to automatically integrate these information sources into a unified digital plant model, it is necessary to develop algorithms for identifying corresponding elements from piping and instrumentation diagrams and 3D CAD models. Regarding this purpose, this section gives an overview of how a process plant is structured and which components it is composed of. In addition, this section describes the piping and instrumentation diagram flow chart, which will be an important part of this approach.

4.3.1 Structure Piping Plant

Piping engineering is a special discipline of mechanical engineering that deals with the design and construction of a pipeline system, the arrangement of the individual process units, and the sequence of elements. Process equipment is typically used for chemical, petrochemical, wastewater, and hydrocarbon feedstocks. Piping, on the other hand, is defined as an arrangement of piping components used to convey and distribute process fluid from one unit to another in a process plant [16]. A process plant is built by a designer according to the requirements of a plant operator. The construction of a plant is done according to some norms and standards, which will not be discussed further.

The process plant or piping plant is described as "the totality of all piping systems in a power or industrial plant including associated intermediate plant structures, accesses, covers, penetrations, insulation, painting, and marking" [16]. Thus, the piping system is the totality of and is composed of the components listed below. To better understand the structure, the piping system is further divided below. The division is made from "large to small" [16]. It should be noted at this point that a process plant is assembled individually. The division says nothing about whether and how often a component occurs in a plant.

The overall piping system consists of several piping systems. These piping systems often combine several pipelines into one functional unit. This means that a piping system, independent of the overall system, is able to operate independently. A pipeline, on the other hand, comprises the connection of system parts that enable the flow medium to be passed on.

As a part of a pipeline, there is the pipeline string or the pipeline section. This is defined as part of a pipeline that, in process plants, begins or ends at a piece of equipment, at a free end, at a branch or bearing, or at the transition to another nominal pressure [16]. During the design process, the designer often needs a connection between two points in the space which is then realized as the pipeline string. In the context of a design change, the correct recognition of a pipeline string gets particular importance as a basic means of automatic recognition of the entire pipeline system.

Pipeline parts are a component of every pipeline. In the overall system, the pipeline parts make up the largest proportion of components and are therefore one of the most important categories. The pipeline components are fastened to the intended position with the help of pipe supports. This pipe supports the transfer of the pipeline loads to the supporting structure that encloses the plant. Due to their quantity, pipe supports are often manufactured as standard parts and, thus, designed by inserting a model from the library. Fittings are pipeline parts with the task of interrupting, throttling, regulating material flows, preventing backflows, or limiting pressures in a system.

A pipeline may be encased in insulation. Insulation is a collective term for cold and heat insulation. In addition, this can also be a simple coating that identifies the pipes according to what kind of material flow is present in the pipes. Insulation can be built as a protective layer of insulating material or a specific cover/shaft for the pipe bundle which hides the pipe. There is also the possibility that the insulation has the function of noise protection, to protect the environment.

Another part of a pipeline system is the equipment which is usually divided into two types. On the one hand, there are the working machines, especially with the tasks of conveying liquids, separating mixtures, mixing or combining individual phases, and reducing screening, or sifting materials. On the other hand, there are power machines, which have the tasks of generating electricity or driving processes. In addition, various appliances can be found in a processing unit. In such reactors physical, chemical as well as biochemical material conversions or heat exchanges take place.

The process control technology controls regulates, and monitors the entire material and energy flow. Thus, the instrumentation and control equipment is an important part of controlling the production process. The task of process control technology is to link the required control components with each other and to establish the flow of information.

Another important component in the entire pipeline system is the storage units where the material can temporarily remain. These can be divided into containers, tanks, silos, or bunkers. Usually, liquids are stored in a container or a tank. The difference between the two storage units is that the container is placed horizontally, and the tank is placed vertically. Silos or bunkers, on the other hand, are "large containers for solids with appropriate discharge devices such as cellular wheels, screw conveyors and turntables" [16].

To ensure that the main process functions as faultlessly as possible, a supply unit is needed. This contains all the necessary support required in a main process, such as water, electricity, or compressed air. In the literature, the supply unit is not explicitly mentioned. Through the interview with the experts from the pipeline construction, the supply unit is identified as a very important factor that is affected by changes during revision/modernization/overhaul.

This overview of the components of a piping system discovers the importance of a fast and seamless design change of the pipeline system in case of revision/modernization/overhaul.

4.3.2 Piping and Instrumentation Diagram (P&ID)

A piping and instrumentation diagram (P&ID) describes, in a schematic drawing format, the sequential flow of liquids, gases, and vapors and how they enter, flow through, and exit the processing facility. By using simplified drawing symbols, to represent various pieces of mechanical equipment, valving, and instrumentation, and specific notes, callouts, and abbreviations, the flow diagram provides the piping designer and other stakeholders with an overall view of the operation of a facility [17].

A P&ID provides a basis of understanding for all people involved in the design, construction, and operation of a process plant, but also the process carried out.

According to DIN EN ISO 10628, the following information is contained in a P&I diagram [18]:

- Flows/quantities of energies/energy carriers, flow path and direction of the energies/energy carriers.
- Type of appliances and machines in the plant.
- Characteristic sizes of appliances and machines.
- Characteristic data of driving machines.
- Arrangement of essential armatures.
- Designation of these armatures.
- Elevation of essential appliances and machines.
- Materials of appliances and machines.
- Designations of the nominal size, pressure rating, material, and design of the pipelines.
- Information on the insulation of appliances, machines, pipelines, and armature.
- Task definition for measurement, control, and regulation.
- Type of important devices for measurement, control, and regulation.

The P&ID is the process flow diagram plus the components of the electrical, measuring, control, and regulating stations. In addition, the relevant parameters such as maximum pressures, nominal pipe size, etc. can be read out in a P&ID. Thus, the flow diagram forms the basis for every project that is related to the mapped plant.

The P&ID takes an important role in this approach. More details are described and explained in the conceptual design in Chap. 6.

4.4 Interoperability Standards for Process Plants

A digital twin is too complex to be the subject of only one standard or initiative. Therefore, many activities are undertaken regarding the standardization of digital twins, even if not explicitly under the title "Digital Twin". Besides the classical standardization organizations like the International Organization for Standardization (ISO) and the International Electrotechnical Commission (IEC), other consortia such as the Industry IoT Consortium (IIT) and the W3C Web of Things (WOT) work on specifications of digital representation of things. The interoperability as a fundamental capability of the digital twin is the key constituent of almost every initiative and approach [19]. An excerpt of the initiatives, the frameworks, and the standards regarding the digital twin in a process plant is listed in Table 4.1.

Under the stewardship of ISO/IEC JTC 1/SC 41 Internet of Things and digital twin Technical Committee, two signature digital twin standards are emerging under development: ISO/IEC AWI 30172 Digital twin—use cases [20], and ISO/IEC AWI 30173 Digital twin—Concepts and terminology [21].

IEC 62832 defines the general principles of the Digital Factory framework (DF framework), which is a set of model elements (DF reference model) and rules for modeling production systems, although it is not called a digital twin. It applies to the

Table 4.1 Initiatives, frameworks, and standards of the digital twin in industrial domains, excerpt

Document	Scope	Body	Reference
ISO/IEC AWI 30172, ISO/IEC AWI 30173	Digital twin—use cases digital twin—concepts and terminology	ISO/ IEC	[20] [21]
IEC 62832	Digital factory framework with the representation of digital factory assets in its center	IEC	[22]
ISO 23247-1(to 4)	Automation systems and integration—digital twin framework for manufacturing	ISO	[23]
IEC 63280 ED1	Automation engineering of modular systems in the process industry—general concept and interfaces	IEC	[24]
IEC PAS 63088	Smart manufacturing—reference architecture model industry 4.0 (RAMI4.0)	IEC	[25]
ISO TS 18101-1	Automation systems and integration—oil and gas interoperability	ISO	[26]
ISO 15926	Integration of life-cycle data for process plants, including oil and gas production facilities	ISO	[27]

three types of production processes (continuous control, batch control, and discrete control) in any industrial sector (for example aeronautic industries, automotive, chemicals, wood) [22].

In 2019, the IEEE Standards Association initiated a project IEEE P2806 that aims to define the system architecture of digital representation for physical objects in factory environments. A similar approach is taken by Digital Twin Manufacturing Framework ISO/AWI 23247 within ISO TC 184/SC4/WG15. This framework enables plug-and-play for twin elements, focusing mainly on the interfaces and functions of digital twins [23].

IEC 63280 for automation engineering of modular systems in the process industry aims the module integration through a simple import of the modular type package into the process control engineering of the production plant [24].

The German Plattform Industrie 4.0 launched Asset Administration Shell as the implementation of the digital twin for smart manufacturing, IEC PAS 63088. This was deepened by partnerships between France, Italy, and Germany [25].

ISO TS 18101-1 provides requirements, specifications, and guidance for an architecture of a supplier-neutral industrial digital ecosystem with a focus on oil and gas interoperability. Digital assets are not considered to be necessarily physical. This document is focused on interoperability requirements for systems that play roles in the secondary business process [26].

The scope of ISO 15926 is "Industrial Automation Systems and Integration of Life-cycle Data for Process Plants, including Oil and Gas Production Facilities" [27]. This is the standard of paramount importance for the acquisition of brownfield process plants because it ensures the standardization of the flow of information between all applications in a process plant during its long-lasting lifecycle. Where many of today's systems require "data cleansing," the aim of ISO 15926 is to map data at the source into the ISO 15926 format, with all due quality assurance before actual storage [28].

ISO 15926 is a further derivate of ontology domains [29] for data modeling and interoperability using the Semantic Web. ISO 15926 is subdivided into eleven parts, each published separately. The representation of the process plant lifecycle is specified by a generic, conceptual data model (described in part 2 of ISO 15926) which is designed to be used in conjunction with reference data (described in part 4), which are standard instances that represent information common to a large number of users, process plants, or both. The support for a specific life-cycle activity depends on the availability of appropriate reference data in conjunction with the data model. To manage a change, the developers of ISO15926 conceptualized a library that will facilitate anyone to add new terms. One of the continuous development tasks of ISO15926 is to generate and extend an industrial-scale reference data library [28].

The data model (part 2) supports all disciplines and lifecycle stages, and information about functional requirements, physical solutions, types of objects, and individual objects and activities. It also defines the rules and constraints on how to use this standard. It is an ontology with the basic axioms such as things, class, and individuals at the top level. At a lower level, it includes subtypes of these axiomatic concepts—such as physical objects and connections. This is a highly generic, 5NF, data model with 201 entity types; it is unconstrained except for disjointness. The constraints must come from the application that produces the data [28].

Part 4 comprises a common set of definitions called reference data library (RDL). While RDL assigns a unique identifier to each definition, each identifier can be used as a serial number for the definition. The RDL contains also individual objects that are important enough that we want to differentiate them from all other objects in the world ("reference individual"). ISO15926 allows the use of a federation of data libraries, where different authorities around the world become respected repositories of certain categories of information. The core of ISO15926 is the data model (Part 2) and the RDL (Part 4), where the core RDL contains the core classes and references individuals, which are arranged in an ontology of subtypes or specializations of the classes in Part 2. Currently, there are almost 20.000 classes in part 4. The magic comes from using Part 2 (the data model) and Part 4 (the dictionary) together to translate data exchanges into the common language of ISO so that it can be translated by business partners at each other end [28] (Fig. 4.6).

In the case of a brownfield process plant, the generation of a digital twin must include a step where the portions of the physical twin (e.g., portions of a point cloud from a scan) are recognized and, thus, replaced by a predefined part from the reference data library. In this way, the recognized object is inserted into the standard design process. The occurrence of "reference individuals" remains an absolute exception. Standardization organizations, companies, and projects can set up their local RDLs in which they define the concepts for which they have responsibility. Examples are ASME B16.11 PIPE ELBOW CLASS, SIEMENS 1MA3133-4NA86, or HC (a project-defined insulation class) [28].

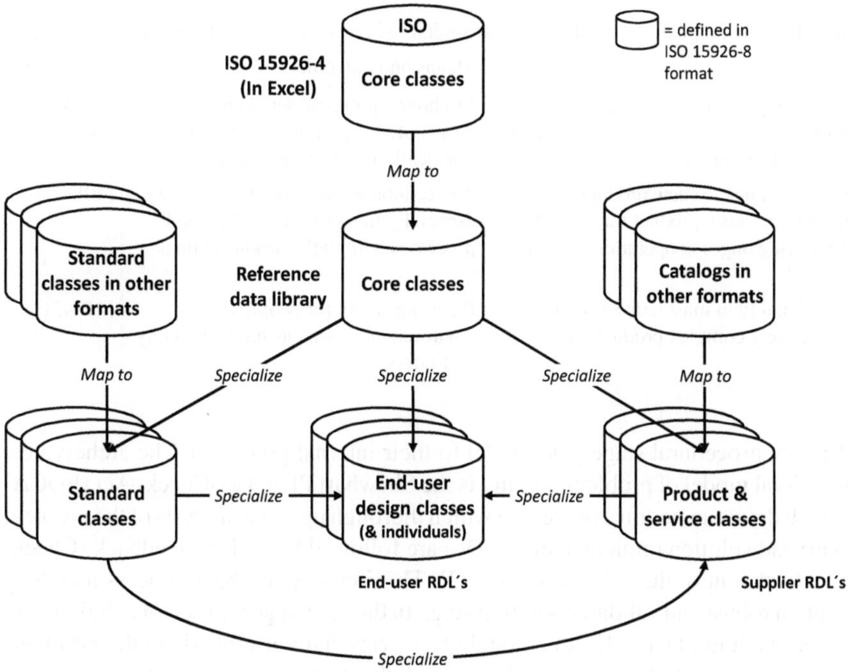

Fig. 4.6 Typical setup for RDL extensions [27]

4.5 Discussion

The primary purpose of the digital twin of a process plant is to seamlessly provide plant information to operators and machines throughout the network of enterprises involved in designing, commissioning, and operating the plant [30]. At the time of commissioning, this information includes design information as well as rules of operation [31]. The lack of information that exists in a brownfield plant needs to be closed accordingly, e.g., by generation and update of a digital twin [32]. It is not claimed that the digital twin offers a perfect solution for all these issues and must be developed in an iterative workflow along the process and plant lifecycle [33]. The idea of scaling projects can be realized in such a way that a benefit is quickly created for all parties involved [34].

Optimized lean processes, e.g., according to the Lean & Six Sigma approach, are a basic prerequisite and starting point for digitalization. Right from the start, the design of the digital twin must follow a multicriteria objective [35], based on three global goals. Within the framework of the multicriteria approach, decision-makers should be enabled to find the best solutions in the respective context in a transparent and flexible manner [34] (Table 4.2).

A comprehensive process modeling must be incorporated in a lifecycle-wide information management strategy and supporting toolchains are required. Most companies

Table 4.2 Global goals for the design of the digital twin in the process industry, derived from [34]

Goal	Means and outcome	References
Design ongoing processes and new product launches with a view to reducing the use of resources	Modular equipment setup and AI-assisted experimentation will drastically reduce time and lab effort	[36, 37] [38, 39]
Reaction times for new product launches (time-to-market), product transfers, but also during ongoing operations can be reduced	The economic benefits will be evident, especially in regulated markets and processes with high working capital	[40]
The digital twin must help to control increasingly complex production processes	Ensuring safety for people and the environment as well as product quality and safety	[41, 42]

adapt the procedural stage-gate model to their internal processes. The archetypical procedural model of problem-solving is the Shewhart Plan–Do–Check–Act (PDCA) cycle. It draws an iterative process in which thorough up-front analysis of the problem (Plan) and solution implementation (Do) are followed by seeking feedback (Check) and adjustment of the solution (Act) [43]. This iterative feedback process may help to obtain robust and validated solutions e.g., in the case of generation of a digital twin of a brownfield plant [44]. It is thought to be especially useful where the problems being solved are ill-defined, complex enough that they cannot be easily grasped, are set in a changing context, and/or in a context where the solution can influence the nature of the problem. With this model, the need for an update of a digital twin can be identified and its execution incrementally tackled [45]. The main value of these models in practice is arguable to assist in communicating methodological insights to a large number of employees (internally and externally), and as such, clarity of exposition may be one of their most important characteristics [43].

The consistency between the 2D information (P&ID) and 3D information from CAD poses a further challenge. If the digital twin is being created from scratch, the consistency between P&ID and the 3D model is ensured in the CAD system for plant design. In the case of a brownfield plant, the 3D model must be created in an alternative way (manually or by object detection) and connected with P&ID manually or by recognition of semantics. This can be done by combining the control loop information from the P&ID with the physical layout from the 3D CAD. A straightforward approach for information integration would be to match tag names from the 2D and 3D information sources to identify the parts of these models that correspond to the same component. However, with industrial design repositories, it cannot be assumed that consistent naming conventions have been enforced to enable this approach [14].

Based on this approach, the digitalization procedure should run in the following 4 steps:

1. Digitize the information to a standard, industrially accepted Industry 4.0 neutral format. This may involve no effort if the designs were made in tools that support these formats. However, industrial plants have lifecycles of several decades, in

which case innovative applications are required to digitalize the legacy design information.
2. Adjust the level of abstraction of the 2D and 3D designs, so that they are at the same level of abstraction.
3. Match the models generated in step 2, to identify the elements in these models that correspond to the same plant component, such as a tank or pump.
4. Use the matches to augment applications relying only on either 2D or 3D information sources. If the instrumentation in the 2D and 3D sources could be matched, it would be possible to automatically identify and connect the I/O interface of the physical and virtual aspects of the digital twin, eventually aiming at the automatic generation of a system that could be considered a fully-fledged digital twin.

However, plant design information still primarily resides in proprietary and tool-specific formats. A few exceptions exist, such as the Proteus XML schema for P&ID exchange, which is supported by several leading P&ID tool vendors, and the PCF (Piping Component File) format for 3D isometrics, supported by leading tool vendors such as Hexagon PPM, Autodesk, and PTC [14].

4.6 Conclusions and Outlook

While the engineering of a process plant is now a thoroughly developed discipline, tried and tested processes, methods, and tools are available for its realization. With a commercially available toolset and customization of low extent, an almost seamless process chain for a long period of the entire lifecycle of a process plant can be realized—in the case of a greenfield plant. Digital asset information management delivers a 'digital twin' of up-to-date information that accurately describes the current condition of the physical asset, rather than how it was originally designed. It directly addresses the challenge of finding information in siloed business systems, making it easy for users to quickly find and aggregate the information they need to make decisions [46].

In the case of a brownfield plant, only the information needed for operation is available to a sufficient extent. However, an information gap exists regarding the asset description, in particular the 3D information. The parameterization of the process components takes place early in the design process of an overhaul. At this stage, the structure of some process areas has already been designed: the main components and the connections between them are known. This information is typically represented in the form of a P&ID. However, the level of detail is still rough, and the physical environment of the unit in question, described by an appropriate 3D model, is missing. This information is crucial, as the design process of the plant consists of multiple subprocesses working in parallel. The required model parameters can be determined with a lab-scale plant, which is successfully down-scaled from the production scale. On the basis of the digital twin, the components can be categorized, and their extraction behavior can be observed for different extraction techniques. Moreover, the possibility to carry out a large number of simulations or different operating points can be used to establish a design space for a complex process [47].

Some of these subprocesses, for example, initial cost and material estimations, require information about the extent of changes before they can continue forward. The more detailed the description of the physical environment, the more exact the estimation for the parameter values, so that the parametrization phase does not create a bottleneck in the project. The timely discovery of estimates for the parameters thus has a significant impact on the entire project schedule, and the quality of the estimates impacts the work of the other design processes [48].

Many aspects of plant engineering are covered by international standards and initiatives which foster collaborative, cross-border engineering. The development of digital twin usage goes hand in hand with the development and use of digital engineering solutions where users stay connected over different physical locations and the within working spaces, using the capabilities of these environments to allow users to interact with each other. The extent of tool-supported design interaction surpasses the human face-to-face exchange using the progress of technology to pass over hurdle after hurdle of needs and goals of collaboration and communication.

The added benefit of the digital twin is its capability to help all stakeholders visualize and promote collaboration and co-creation of policy and sustainability improvement solutions among stakeholders, who often come from different technical backgrounds [49]. The development of advanced model-tool-product systems needs the development of holistic engineering approaches. These approaches can offer the possibility of the design of intelligent self-sustainable models and intelligent self-sustainable products. It could be realized by a full and meaningful quantification of the performance of a digital twin [50]. The self-evolving nature of the digital twin can be better implemented if the digital twin can assess its own performance.

References

1. Sutton I (2015) Plant design and operations. Elsevier, Waltham, p 2015
2. Selcuk Agca H, Cotone G (2019) Introduction to process plant projects. CRC Press, Boca Raton
3. Nikjow MA, Liang L, Qi X, Sepasgozar S (2021) Engineering procurement construction in the context of belt and road infrastructure projects in West Asia: a SWOT analysis. J Risk Financ Manage 14:92. https://doi.org/10.3390/jrfm14030092
4. Basu S, Debnath AK (2019) Power plant instrumentation and control handbook: a guide to thermal power plants. Elsevier, London
5. Pistikopoulos EN, Barbosa-Povoa A, Lee JH, Misener R, Mitsos A, Reklaitis GV, Venkatasubramanian V, You F, Gani R (2021) Process systems engineering—the generation next? Comput Chem Eng 147:107252. https://doi.org/10.1016/j.compchemeng.2021.107252
6. Thomas CE (2015) Process technology equipment and systems, 4th edn. Cengage Learning, Stamford
7. Akbulut A, Arslan O, Arat H, Erbaş O (2021) Important aspects for the planning of biogas energy plants: Malatya case study. Case Stud Therm Eng 26:101076. https://doi.org/10.1016/j.csite.2021.101076

8. Leidich J, Dagner T, Robl P (2022) Optimized planning of the integration of a reference plant into existing brownfield environments based on an entity model. In: DS 119: proceedings of the 33rd symposium design for X (DFX2022). https://doi.org/10.35199/dfx2022.14
9. Lock D, Wagner R (eds) (2018) The handbook of project portfolio management. Routledge, London. https://doi.org/10.4324/9781315206592
10. Houson I (2011) Process understanding—for scale-up and manufacture of active ingredients. Wiley, Hoboken
11. NN (2020) Usage view "seamless and dynamic engineering of plants". Federal Ministry for Economic Affairs and Energy (BMWi), Berlin. https://www.plattform-i40.de/IP/Redaktion/EN/Downloads/Publikation/usage-view-seamless-and-dynamic-engineering-of-plantsx.pdf. Accessed 20 Oct 2022
12. Löwen U, El Sakka F, Schertl A, Fay A (2020) A systematic approach how to build, maintain and use an integrated plant model. Automatisierungstechnik 68(6):423–434. https://doi.org/10.1515/auto-2019-0097
13. NN (2022) The industrial internet reference architecture technical report. https://www.iiconsortium.org/IIRA.htm, Accessed 20 Oct 2022
14. Sierla S, Azangoo M, Fay A, Vyatkin V, Papakonstantinou N (2020) Integrating 2D and 3D digital plant information towards automatic generation of digital twins. In: IEEE 29th international symposium on industrial electronics (ISIE). https://doi.org/10.1109/ISIE45063.2020.9152371
15. Helmus FP (2003) Anlagenplanung: Von der Anfrage bis zur Abnahme. Wiley-VCH Verlag GmbH & Co, KGaA, Weinheim
16. Wossog G (2008) Rohrleitungsbau, 3rd edn. Vulkan-Verlag, Essen
17. Parisher RA, Rhea RA (2021) Pipe drafting and design, 4th edn. Elsevier, Amsterdam. https://doi.org/10.1016/C2019-0-01022-9
18. Bindel T, Hofmann D (2016) Aufbau des R&I-Fließschemas nach DIN 19227 bzw. DIN EN 62424. Springer-Verlag, Berlin. https://doi.org/10.1097/978-3-658-15559-9
19. NN (2022) Digital twin consortium, publications. https://www.digitaltwinconsortium.org/publications/. Accessed 30 Nov 2022
20. NN (2022) ISO/IEC AWI 30172, digital twin—use cases. https://www.iso.org/standard/81578.html. Accessed 30 Nov 2022
21. NN (2022) ISO/IEC AWI 30173, digital twin—concepts and terminology. https://www.iso.org/standard/81442.html. Accessed 30 Nov 2022
22. NN (2020) IEC 62832-1:2020, industrial-process measurement, control and automation—digital factory framework—part 1: general principles. https://standards.iteh.ai/catalog/standards/iec/cfef54c6-c080-4e1d-980a-a7d53a784409/iec-62832-1-2020. Accessed 30 Nov 2022
23. NN (2021) ISO 23247-1:2021, automation systems and integration—digital twin framework for manufacturing—part 1: overview and general principles. https://www.iso.org/standard/75066.html. Accessed 30 Nov 2022
24. NN (2022) IEC 63280 ED1, automation engineering of modular systems in the process industry—general concept and interfaces, IEC. https://genorma.com/en/project/show/iec:proj:103501. Accessed 30 Nov 2022
25. NN (2017) EC PAS 63088:2017, smart manufacturing—reference architecture model industry 4.0 (RAMI4.0). https://www.vde-verlag.de/iec-normen/224330/iec-pas-63088-2017.html. Accessed 30 Nov 2022
26. NN (2019) ISO/TS 18101-1:2019, automation systems and integration—oil and gas interoperability—part 1: overview and fundamental principles. https://www.iso.org/standard/68521.html. Accessed 30 Nov 2022
27. NN (2018) ISO 15926—integration of life-cycle data for process plants, including oil and gas production facilities. https://15926.org/home/. Accessed 30 Nov 2022
28. Yacout S, Ebrahimipour V (2015) ISO 15926. In: Ebrahimipour V, Yacout S (eds) Ontology modeling in physical asset integrity management. Springer, Cham. https://doi.org/10.1007/978-3-319-15326-1_1

29. da Silva Serapião Leal G, Guédria W, Panetto H (2019) An ontology for interoperability assessment: a systemic approach. J Ind Inf Integr 16:100100. https://doi.org/10.1016/j.jii.2019.07.001
30. Dal Pont JP, Azzaro-Pantel C (2014) New approaches to the process industries: the manufacturing plant of the future. Wiley, Hoboken
31. Moreno-Garcia CF, Elyan E (2019) Digitisation of assets from the oil and gas industry: challenges and opportunities. In: Proceedings of 2019 international conference on document analysis and recognition workshops, vol 7. IEEE, pp 2–5. https://doi.org/10.1109/ICDARW.2019.60122
32. Stjepandić J, Sommer M, Denkena B (2022) DigiTwin: an approach for production process optimization in a built environment. Springer International Publishing, Switzerland. https://doi.org/10.1007/978-3-030-77539-1
33. Fukuda S, Lulić Z, Stjepandić J (2013) FDMU-functional spatial experience beyond DMU? In: Proceedings of the 20th ISPE international conference on concurrent engineering, CE 2013. IOS Press, Amsterdam, pp 431–440. https://doi.org/10.3233/978-1-61499-302-5-431
34. Bamberg A, Urbas L, Bröcker S, Bortz M, Kockmann N (2021) The digital twin—your ingenious companion for process engineering and smart production. Chem Eng Technol 44:954–961. https://doi.org/10.1002/ceat.202000562
35. Adamenko D, Kunnen S, Pluhnau R, Loibl A, Nagarajah A (2020) Review and comparison of the methods of designing the digital twin. Procedia CIRP 91:27–32. https://doi.org/10.1016/j.procir.2020.02.146
36. Kockmann N, Bittorf L, Krieger W, Reichmann F, Schmalenberg M, Soboll S (2018) Smart equipment—a perspective paper. Chem Ing Tec 90(11):1806–1822. https://doi.org/10.1002/cite.201800020
37. Hohmann L, Kössl K, Kockmann N, Schembecker G, Bramsiepe C (2017) Modules in process industry—a life cycle definition. Chem Eng Process 111:115–126. https://doi.org/10.1016/j.cep.2016.09.017
38. Fath V, Kockmann N, Otto J, Röder T (2020) Self-optimising processes and real-time-optimisation of organic syntheses in a microreactor system using Nelder-Mead and design of experiments. React Chem Eng 5(7):1281–1299. https://doi.org/10.1039/D0RE00081G
39. Fath V, Lau P, Greve C, Kockmann N, Röder T (2020) Efficient kinetic data acquisition and model prediction: continuous flow microreactors, inline Fourier transform infrared spectroscopy, and self-modeling curve resolution. Org Process Res Dev 24(10):1955–1968. https://doi.org/10.1021/acs.oprd.0c00037
40. Bieringer T, Buchholz S, Kockmann N (2013) Future production concepts in the chemical industry: modular—small-scale—continuous. Chem Eng Technol 36:900–910. https://doi.org/10.1002/ceat.201200631
41. Kockmann N, Thenée P, Fleischer-Trebes C, Laudadio G, Noël T (2017) Safety assessment in development and operation of modular continuous-flow processes. React Chem Eng 2:258–328. https://doi.org/10.1039/C7RE00021A
42. Lee J, Cameron I, Hassall M (2019) Improving process safety: what roles for digitalization and industry 4.0? Process Saf Environ Prot 132:325–339. https://doi.org/10.1016/j.psep.2019.10.021
43. Wynn DC, Clarkson PJ (2018) Process models in design and development. Res Eng Design 29:161–202. https://doi.org/10.1007/s00163-017-0262-7
44. Grau M, Korol W, Lützenberger J, Stjepandić J (2021) Automated generation of a digital twin of a process plant by using 3D scan and artificial intelligence. Adv Transdisciplinary Eng 16:93–102. https://doi.org/10.3233/ATDE210087
45. Kremer P, Lützenberger J, Müller F, Stjepandić J (2022) An approach for the incremental update of a digital twin of a process plant. Adv Transdisciplinary Eng 28:310–319. https://doi.org/10.3233/ATDE220660
46. NN (2022) AVEVA™ asset information management discovery. https://www.aveva.com/en/solutions/operations/asset-visualization/. Accessed 22 Nov 2022

47. Uhlenbrock L, Jensch C, Tegtmeier M, Strube J (2020) Digital twin for extraction process design and operation. Processes 8(7):866. https://doi.org/10.3390/pr8070866
48. Rantala M, Niemistö H, Karhela T, Sierla S, Vyatkin V (2019) Applying graph matching techniques to enhance reuse of plant design information. Comput Ind 107(2019):81–98. https://doi.org/10.1016/j.compind.2019.01.005
49. Kaewunruen S, Lian Q (2019) Digital twin aided sustainability-based lifecycle management for railway turnout systems. J Cleaner Prod 228:1537e1551. https://doi.org/10.1016/j.jclepro.2019.04.156
50. Sharma A, Kosasih E, Zhang J, Brintrup A, Calinescu A (2022) Digital twins: state of the art theory and practice, challenges, and open research questions. J Ind Inf Integr 30:100383. https://doi.org/10.1016/j.jii.2022.100383

Chapter 5
Business Case for Digital Twin of a Process Plant

Abstract The development of new offerings (products or services) belongs to a value-creation process. Such a process requires conceptual design and practical implementation. Hereby, the focus relies on how to create value for the customer. An accordingly defined business model is the key point. It describes the interdependencies between the product/service and its development. For that purpose, the multiple value dimensions of the product or service are explored as the constituents of the business planning. The concept of the digital twin provides various benefits for customers. Especially for stakeholders in plant industries, that deal with complex assets, a digital twin providing 3D and spatial information respectively is of value. The impact of this concept could not be exploited so far since the costs for the generation of such digital twins are enormous, especially for brownfield plants. For that reason, a digital twin has only been generated for projects with an according return on investment. In that sense, the potential of digital twins is limited to a few business cases in the process industry. An automated generation of digital twins or at least a semi-automation reduces the costs to a reasonable level. Digital twins can be kept up-to-date with reasonable effort and cost this way. That enables stakeholders to exploit digital twins also for business cases with a lower return on investment such as maintenance activities. According to the concept of an automated generation of digital twins with spatial information, current offerings in the market have been analyzed and reflected by applying the buyer utility map methodology. Further, the new offering and its advantages and benefits have been added to differentiate between them. To shape a suitable business model according to the new offering, the approach of the business canvas has been exploited. The various fields to consider by developing the service solution, requirements from the target groups, the value proposition, and further have been defined that way. Finally, the customer's view of the concept of digital twin and use cases of the concept of digital twin including spatial information is presented and underlines the need for the mentioned service solution.

Keywords Digital Twin · Product Lifecycle Management · Business model · Business canvas · Buyer utility map · Object recognition from point cloud · Process industry

J. Stjepandić et al., *Generation and Update of a Digital Twin in a Process Plant*,
https://doi.org/10.1007/978-3-031-47316-6_5

5.1 Introduction

The development of new products or services belongs to a value-creation process. It ensures that customer needs or requirements are considered and addressed within new offerings. Here, the value is seen as an advantage for the customers from the customer's perspective. Value as defined in relation to process reflects how the goal-pursuit activity itself is experienced. In consumer contexts, this may often reflect decision-making processes. Therefore, new solutions need to address and solve problems and support current processes e.g., by making them more efficient or providing additional benefits to the customer and its business case [1]. The value defined in terms of outcomes reflects the nature of the end state—what people want, or the end goal. Historically, an outcome is valued to the extent that it is useful or satisfies some need, or produces pleasure and not pain [2].

Pre-condition is a deep understanding of the existing cases, the market, and the own solution. Value creation processes are helping to understand the strengths of the own solution, their differentiation from existing products or services, and how they address customer needs [2]. They basically analyze the relevance of solutions and their right to exist and finally hold one's ground in the market in three stages. The interrelationship between risk and opportunity, strategy and value creation, and the impact of external contextual factors is analyzed first. Secondly, the role of corporate reporting and reporting processes and governance in shaping those linkages to create value is the following. Lastly, the role of individual actors and leadership in tempering the links between risk and opportunity, strategy and value creation, and the interplay with corporate governance, reporting, and external contextual factors is analyzed [3]. Even though such analysis also provides arguments for better marketing and convincing the customer to buy a product, it is not the purpose of value creation processes.

The value creation process can be applied through various methods, for instance, boundary spanning, experimentation, practical tools, and guidelines. These methods support the reflection of business cases or products/services from various levels and perspectives [4]. A deep understanding of customer expectations and their own solution can be achieved this way. Additionally, the market can be explored and the target groups identified [5]. In this sense, products and/or services need to be shaped in different ways if they are designed for end-customers or business customers and the market (B2B or B2C) respectively [6].

Sustainable business models are based on a triple-bottom-line approach and consider a wide range of stakeholder interests, including the environment and society. In a comprehensive comparison, four different types of sustainable business models could be distinguished. These business models are considered sustainable because, compared to traditional business models, they either (1) imply improvements towards efficiency, (2) are based on new ways to make the business sustainable, (3) have a stronger orientation towards society and/or the environment, or (4) are "born sustainable." [4].

Conversely, sustainable value creation requires (1) a stakeholder-responsive definition and understanding of value; (2) a systems approach that includes spatial and temporal aspects to identify the recipients of value; (3) a relational interpretation of and collaborative approach to value co-creation; and (4) measures of total value creation that consider power relationships and value capture patterns that occur among stakeholders [7].

There are some barriers to overcome on the way to business model innovation. Organizational leadership, a clear vision, management support, organizational structure, incentives, resources, and change processes were said to contribute to business model innovation. On top of this, for sustainable business model innovation, a sustainability vision would be required, as well as performance management, metrics focused on sustainability, personal leadership, sustainability values, and collaboration with stakeholders [8].

The dynamic adaptation to customer needs remains in line with the evolutionary nature of the market. It covers three key aspects—customer sensitivity, the existence of communication channels and dialogue and the transformation of knowledge exchanged into specific measures for creating customer value. Furthermore, this co-existence not only leads to the creation of customer value but also to global improvement in the firm's competitiveness, understood to be a multi-dimensional construct measured in terms of profitability, post-optimization, and technology use [9]. These aspects can be extrapolated to supply chains [10].

Despite the emphasis on the collaborative nature of value creation, the literature provides scarce elaboration on the joint activities, principles, and phases of value co-creation for digital servitization. A proposed micro-service innovation approach is built on three principles: incremental micro-service investments, sprint-based micro-service development, and micro-service learning by doing. These principles are implemented in digital servitization through an iterative five-phase agile co-creation process [11].

Moving beyond value creation in individual companies, firms have integrated customers, partners, and stakeholders in a mutual value co-creation process which yields the platform to develop and share applications in an ecosystem. A case study in the context of emerging Internet-of-Things (IoT) platforms shows that B2B platforms follow three standardized value co-creation practices. The platform encourages the supply side through the (1) integration of complementary assets, the demand side through (2) ensuring platform readiness and connects both processes by (3) servitization through application enablement [12].

Different digital platform ecosystems are compared according to three core building blocks: (1) platform ownership, (2) value-creating mechanisms, and (3) complimentary autonomy. We conclude by giving an outlook on four overarching research areas that connect the building blocks: (1) technical properties and value creation; (2) complimentary interaction with the ecosystem; (3) value capture; and (4) the make-or-join decision in digital platform ecosystems [13].

The digital servicescape with artificial intelligence (AI) capabilities generates opportunities for multiple actors to participate in service ecosystems, creating an environment that is both competitive and collaborative. Essentially, the AI context of digital servitization may create an interdependency where the actors integrate their skills and knowledge of AI technology and consumer data and need to offer superior AI-enabled value propositions to the consumer [14].

In addition, an understanding of the own position in the market as an offering party is important. First, it needs to be clear if the developed solution will be brought into a market with already existing competitors, the so-called red ocean. Here, the price is one of the most relevant criteria to compete with other companies. On the other hand, a solution can open a new market. In the so-called blue ocean, no competitors are available or the advantages and benefits provided by the developed solution are outstanding so that already existing solutions are not seen as competitors [15].

The structure of this chapter reflects these aims. In the following Sect. 5.2, the definition of offering analysis and its supporting concept is briefly introduced. Conceptual definition of the new business model is discussed in Sect. 5.3. Subsequently, value creation and business planning are drawn in Sect. 5.4. Section 5.5 highlights the customers' view of the digital twin. Customers' demand for a digital twin is presented in Sect. 5.6. Finally, an outlook is given in Sect. 5.7.

5.2 Offerings Analysis

In order to understand the current offerings in the market the methodology of the buyer utility map can be applied, enabling an evaluation of the offerings from the customer's perspective. Two axis span a matrix to visualize the customer's experiences in various stages. The horizontal one considers the different stages of the buyer's experiences:

1. Purchase
2. Delivery
3. Use
4. Supplements
5. Maintenance
6. Disposal.

According to an offering, the customer is going through different stages. It starts with the ordering of a product or service. After getting it delivered, the solution can be used. For some products/solutions supplements are needed. Over time, maintenance might become necessary and also disposal might be of relevance. While a customer is going through the mentioned stages, experiences on various levels are made. The vertical axis is reflecting the different utility levers of a customer [15]:

1. Customer productivity
2. Simplicity
3. Convenience
4. Risk
5. Fun and image
6. Environmental friendliness.

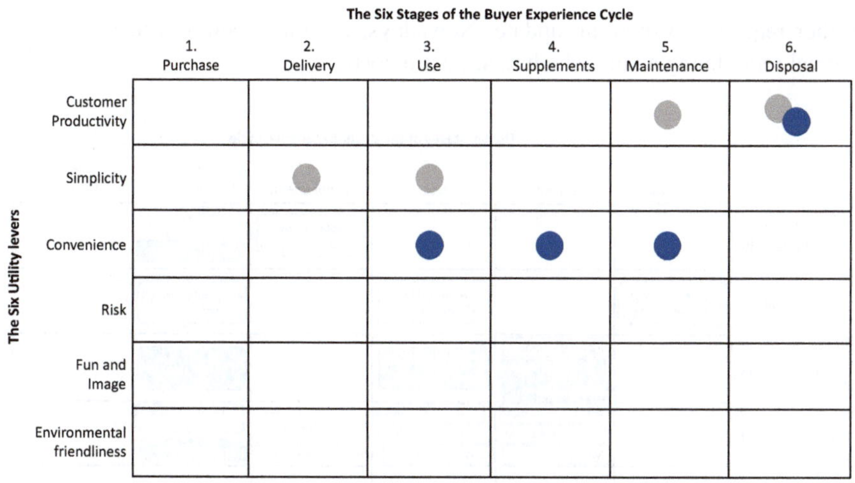

Fig. 5.1 The buyer utility matrix schema [15]

Figure 5.1 shows the buyer's utility map exemplarily. The grey dots mark the cells where other parties in the market already provide solutions. Here, it is not distinguished if this solution comes with benefits or also with disadvantages. In comparison, the blue dots mark cells addressed by a newly developed solution as blue ocean opportunities.

The method of buyer's utility mapping has been applied for the actual use case of generating a 3D representation of an existing plant:

An operator of a process plant is planning to modernize its plant. Some parts of the plant are kept, while others need to be replaced in order to increase productivity. As the basis for this planning, the current as-is state needs to be digitized into CAD models. A planning office is assigned to carry out the as-is documentation in terms of remodeling the plant. The modeling is processed based on a point cloud generated by a 3D laser scanner. It represents the as-is state. During the remodeling process, in the first step, the engineers are placing the equipment, followed by instruments and piping system.

Figure 5.2 shows the buyer utility matrix with regard to the scenario mentioned above. First, the current service offers have been captured and reflected (grey post-its). Nowadays the remodeling must be done manually and is often carried out in countries with a lower salary level as in Europe. Here, many employees are not sufficiently educated, which leads to a reduction in quality. In consequence, the customer needs to double-check the models with assistance from experts with specific knowledge. Oftentimes, these models need to be reworked.

Manual remodeling is time-consuming and this way expensive work. Therefore, it is only to be done for business cases that provide a return on investment accordingly. Modernization is one of them. For other business cases, such as maintenance, a 3D digital twin is not applicable due to the high costs. The business case of maintenance further requires an ongoing update. Nowadays, it would require a new scan and manual remodeling again, which raises extra costs.

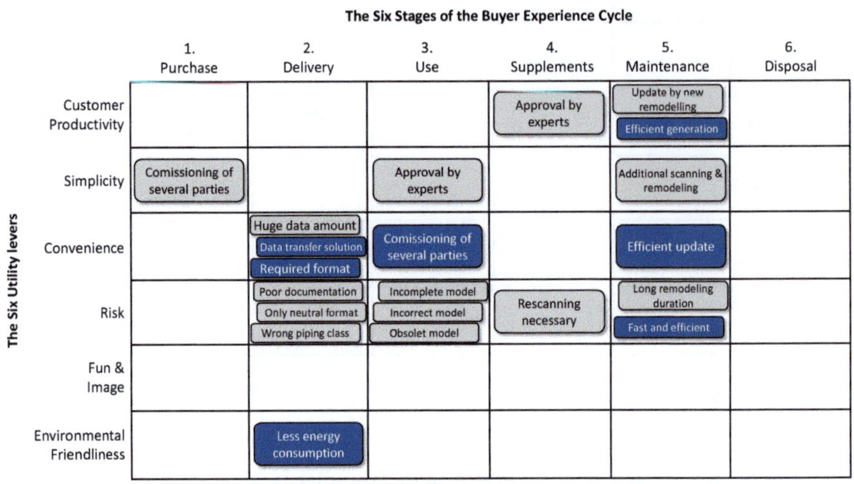

Fig. 5.2 The 3DigitalTwin buyer utility map, derived from [15]

In the process of digitizing an existing plant, there are often several stakeholders involved. In the first step, the actual state of a plant needs to be captured. This can be achieved by a 3D laser scan. The scanning is often carried out by an additional company not carrying out the remodeling or planning. Therefore, the client needs to engage several companies in the worst case. In addition, due to the complexity and size of plants in the process industry, the amount of data is high. Transferring the data (point clouds and CAD models/files) demands a data transfer solution that enables the transfer of data sizes of terabytes. For that reason, nowadays hard disks are often sent by mail.

If the remodeled data has been received, it needs to be double-checked by experts to ensure that the data quality is sufficient. If not, a correction of the data is necessary. Also, additional laser scans might become necessary if relevant parts haven't been captured in the first run. Coordinating the different parties leads to a certain degree of additional effort and complexity.

The lack of quality can also affect the documentation. It needs to be ensured, that the redesigned CAD models are corresponding to the according piping and instrumentation diagrams (P&ID). Even though the P&IDs are the source of information on which components are installed and therefore are compared to the as-is state or CAD models respectively, it often lacks in terms of documentation. If the documentation is incomplete or lacks in terms of quality, tracking and tracing of deviations is not possible on a reliable basis.

While the grey post-its represent the characteristics of offerings available on the market, the blue post-its are related to the service of an automated generation of a digital twin. Some of the characteristics are overlapping. Since the new service requires point clouds as input, these require laser scans beforehand, which need to be carried out by additional companies. In this sense, the simplicity of purchase is comparable to already available offerings.

In both cases, huge data amounts need to be transferred during delivery. However, the new offering benefits from already existing infrastructure for data transfer. The accompanying solution enables customers to transfer data of unlimited size, encrypted and worldwide. A physical sending of storage discs is not necessary. That increases convenience significantly.

Another advantage with regard to delivery is the provision in the required format. While other offerings prefer the delivery in neutral formats or only limited to a few native formats, the envisaged solution benefits from experiences concerning CAD conversion. This enables delivery in the required format (native or neutral).

Even though environmental friendliness is not directly dedicated to delivery, the process of remodeling in a manual manner consumes a lot of energy. A decent amount of people work for several months at high-performance PCs to remodel from the point cloud. In comparison, the envisaged solution is based on artificial intelligence (AI). Of course, spending energy beforehand is necessary to train artificial intelligence. However, the object recognition process will run within days or hours. This way the energy consumption can be reduced in comparison to manual modeling.

Before the remodeled CAD models can be used for the planning, the models need to be checked by experts. The trained AI is improving with each trained point cloud, however due to manifold conditions (quality of scan, weather, dust, etc.) misinterpretation is possible. To ensure that planning is not carried out on basis of an inconsistent model, an expert needs to double-check and, in case of deviations, correct the model.

Nevertheless, the new service solution increases the convenience for the customer significantly. Even if the customer has to complete and double-check the model, the main work is done in an automated manner and provided in the required format. The solution enables an economical generation of a digital twin with spatial information, which can also be used for business cases with a lower return on investment, as mentioned above.

Even though the accuracy increases over time, due to the dependencies of the scan quality and frame conditions, a risk for incompleteness and incorrectness remains. It cannot be ensured that all components have been identified correctly. To compensate for this possible inconsistency, some kind of inconsistency report might be delivered as well. That way documentation can lower this risk.

As already mentioned, the models need to be checked before they can be used for their purpose. The involvement of additional experts for approval can be seen as a supplement to enable the use of the models. A double-check by experts as quality assurance will be necessary also for the new offering but is expected to be less burdensome.

If inconsistencies are identified, these need to be corrected. In some cases, this can be done easily by engineers. In other cases, the point cloud is not sufficient to determine the specific components. That requires further scanning of the plant or at least of the areas of interest. It might be necessary for already conventional and the new offering.

A critical point for the success of digital twins is maintainability. Since existing offerings are based on manual work, they are costly and time-consuming. The new offering of automated generation of a 3D digital twin is an efficient and economical way of generating a digital twin with spatial information (CAD models). Changes can be captured by a laser scan and incorporated efficiently. In this way, thc new offering enables convenient maintenance for customers. It is faster and cheaper.

5.3 Conceptual Definition of the New Business Model

The identification of currently available offerings and discovering the blue oceans was the first step and is described above. In order to define a suitable offering and a business model respectively the approach of the business canvas has been applied. A business canvas is a visual representation of nine core components (Fig. 5.3) [16].

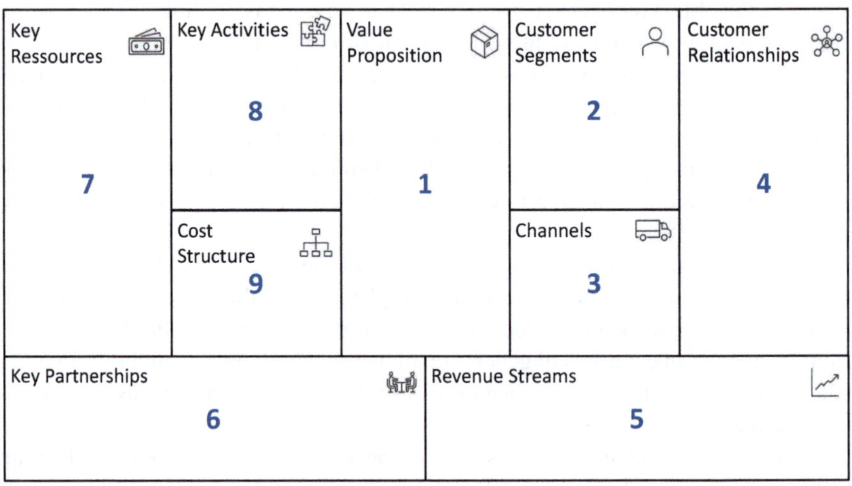

Fig. 5.3 The business model canvas [16]

The numbers in the different fields (Fig. 5.3) reflect its consecutive steps. To add content to the different topics or fields, an expert team sticks post-its to the canvas during a workshop. Starting at the value proposition, the different areas are gone through as follows:

1. **Value proposition**

Here, the offering to the customer is described. It is important to be clear about the benefits for the customer and which advantages are provided. It must be clear why the customer is willing to go for this specific offer.

2. **Customer segments**
   The different target groups or potential customers are listed here. This area reflects to whom the offering is shaped and which customers are addressed by the offering.
3. **Channels**
   When the target group and the offering are defined, the channel to the customer or the access to the offering respectively needs to be defined. This area describes where customers get access to information, marketing channels, and how a customer can place an order.
4. **Customer relationship**
   After the first contact, the relationship with the customers needs to be maintained. According to the offering, different relations can be established, from self-service to personal contact.
5. **Revenue Streams**
   Depending on the offering, various order models might make sense. For only a few buyers, a one-time sales model is suitable. If the customer will order regularly, recurring subscription models might make sense. Also, a license model can be suitable.
6. **Key partnerships**
   In some cases, collaboration with other companies or partners is necessary or at least beneficial. This can range from suppliers of specific materials to development partners that are needed to implement a solution.
7. **Key resources**
   Besides strategic partnerships also the resources needed must be considered. Here, an overview of potential investments or costs, and key roles in terms of persons, patents, trademarks, or infrastructure can be critical for the success of a solution.
8. **Key activities**
   To keep an offering running, various activities are necessary. The most relevant or crucial activities should be listed in this area of the business canvas. Here, the range can also vary from optimization and processing of the offering to customer relations and gaining new customers.
9. **Cost structure**
   Each offer also leads to costs/expenses that arise to be able to provide the offer. This starts with costs for personnel involved, infrastructure costs, licenses, materials, etc.

The abovementioned provides a brief overview of how a business canvas can be filled and what needs to be considered. This way a clear understanding and definition of the business model can be achieved.

5.4 Value Creation and Business Planning

As mentioned, the business canvas methodology can be applied in order to define a business model according to an offering. This method has been applied to detail the offering and to define the business model of a new service providing an automated generation of a digital twin. Within a workshop with experts from the different domains involved, the canvas has been elaborated (Fig. 5.4) according to the methodology described in Sect. 5.2.

The **value proposition** is an automated generation of a digital twin covering 3D or spatial information of a brownfield plant. By exploiting artificial intelligence, object recognition within the point cloud is achieved. That enables a faster and cheaper generation of 3D digital twins in comparison to the offerings of manual remodeling currently available on the market. The result will be provided in the format the customer requires. Neutral formats such as STEP or native formats according to the application in use are possible. By applying the customer's specification in terms of the piping class, an intelligent model can be generated.

With the offering, different **customer segments** are addressed. In the branch of process industry so-called EPCs (Engineering, Procurement, and Construction) are assigned by operators to coordinate and carry out needed tasks from planning to construction of a process plant. They are also in charge when it comes to the modernization of an already existing plant. For the planning of the new parts, a CAD model representing the as-is state of the specific brownfield plant is needed as the basis. Additionally, operators are interested in accurate as-is documentation of their assets. However, the generation of a 3D digital twin is cost-intensive so far and 3D representations are often not applied due to inefficiency. The automated generation of a 3D digital twin also enables the use of those applications, a 3D as-is representation

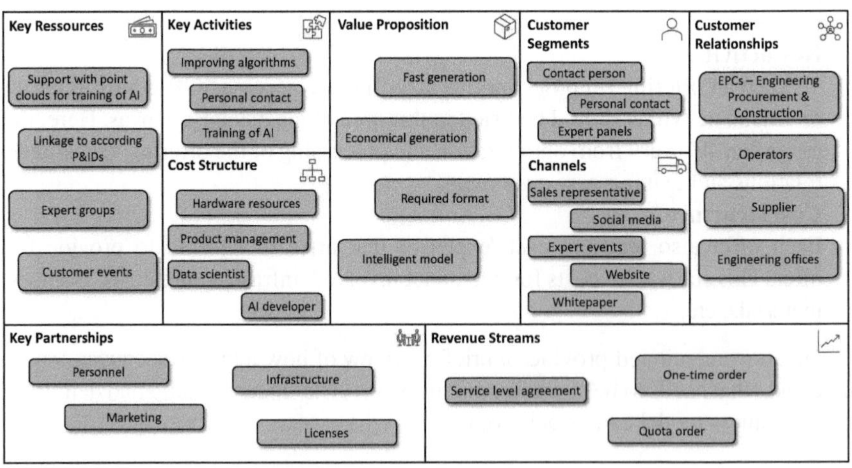

Fig. 5.4 The business model canvas, derived from [16]

has been too expensive until now. Further, engineering offices or service providers and suppliers, e.g., for maintenance activities, can benefit from such digital twins to achieve a better and more precise basis for the planning of their tasks.

The customer groups can be reached through various **channels**. These ensure that information according to the offering is provided and access to the offering is given. Through the company website, basic information and contact to a sales representative is given. Here, also a brochure with basic information about the offering can be downloaded. In addition, marketing campaigns through social media platforms are applied. More detailed information is provided in white papers. They reflect a deep understanding with regard to the needs of the customer groups. Direct access to target groups can be reached through expert events by giving a presentation and demonstration of the service offering. Once customers are reached and are happy with the result of the service word-to-mouth combines recommendation and customer reference based on customer satisfaction, which is probably the most valuable marketing.

When the first contact with the customer is made the **customer relation** needs to be maintained. Since the offering is no self-service solution, direct and personal contact is provided. This way the customer can place orders or gets information. In case of complaints, these can be raised on a personal level and it can be taken care of immediately. Furthermore, expert panels on a regular basis are touch points to interact with the customers and prospects from the target groups and to strengthen the personal relationship with the customers.

In terms of ordering models, three **revenue streams** are defined depending on the frequencies customers are ordering the service. Customers with a selective need can go for a one-time order according to their needs. If the frequency is on a regular level, a recurring subscription is more suitable. Here, it can be distinguished between a quota order or a service level agreement. The quota order offers the customer the possibility to get a discount on the ordered services while the service can be ordered on demand. A service level agreement ensures the customer prioritized processing of the order.

To ensure sufficient implementation of the service, **key partners** are needed. In a first step, artificial intelligence, which enables object recognition from point cloud, needs to be trained. Therefore, point clouds of plants from the process industry are prepared and used. Access to the point clouds is enabled by key partners who support the new service solution. Over time, the point clouds of customers replace these key partners since they improve the capabilities of the AI automatically while they are processed. In addition to the generation of 3D models, partners with additional competencies can expand the offering. In this sense, the linkage of the CAD models of the piping system of a plant can be linked to the according P&IDs. The pre-condition is their availability digitally.

The **key resources** are dedicated to infrastructure and personnel. To apply the object recognition process in an efficient manner, specific hardware is needed. The AI algorithms are shaped according to a specific hardware architecture providing high-performance capabilities. Without suitable hardware, an efficient application of the service solution is not possible and an offering with reasonable conditions (elapse time, recognition accuracy, pricing) cannot be achieved.

There are two critical roles, developing and training the AI. On the one hand, an AI developer needs to implement the algorithms. He or she is in charge of implementing, optimizing/improving, and training the AI. A prerequisite for sufficient training is accurate data science. In this sense, the second important role is the data scientist. He or she is responsible for the preparation of the point clouds that are used for the training. These datasets need to be prepared in the most accurate manner to ensure optimal training results and the generation of a robust data model of the AI.

In terms of management, the product owner and product manager are reflecting the key roles. Both can be seen as sparring roles. While the product owner is in charge of leading the development and in this sense the internal organization, the product manager is responsible for the administrative organization and contact to the target groups. The needs and expectations of the user groups are translated into technical requirements by the product manager. The product owner integrates them on an architectural level into the solution and is responsible for the implementation.

The mentioned key roles are dedicated to **key activities** in order to get the service solution running, improving, and expanding, and to get in contact with the customers. In this context, three main activities have been discovered. Firstly, AI algorithms need to be implemented and improved to ensure an efficient and economical offering. This activity is closely related to the training of AI, which is another major task. Generating a reliable and robust data model is mandatory for the success of the offering. The third key activity is personal contact with customers. This is important in two manners. Firstly, by introducing the solution to customers and experts, feedback can be collected and the solution can be designed according to customer needs. Secondly, by introducing the solution to expert groups and customers, awareness of the solution is ensured and customers can be won.

Finally, the **cost structure** is crucial. Four main investments have been identified to provide the solution. The costs of the personnel involved are one of them. Further, the necessary infrastructure for the development, training, and running of the AI as well as equipment for data science is needed. Here, also specific software is needed, which raises the cost of licenses. Finally, costs for visibility and feedback concluded as marketing costs appear. They are necessary to provide information to target groups, participate in events, get feedback, and ensure visibility.

5.5 Customers' View to the Digital Twin

First of all, the digital twin induces an investment of a significant extent for a user company. Furthermore, the implementation and the introduction of a complex system, such as a digital twin, takes a long period which ties up the resources [17]. Moreover, like each software, the digital twin underlies the updates on a regular basis which usually disturbs the daily operation. Appropriate staff is necessary to keep the digital twin running. These are the cost drivers which may affect the customers' decision to introduce and use the digital twin. Therefore, the process of definition and exploitation of the digital twin must consider the customers' needs and transform those into explicit requirements, as described in Chap. 2, which should be fulfilled by an appropriate service [18].

Here, we will draw the typical customers' thoughts and expectations related to the digital twin and derive the customers' demand for the digital twin. It must be presented in an understandable form to facilitate the customers' decision. Finally, it should encompass the possible approaches to creating a digital twin and its use cases and be justified primarily by value streams resulting from cost savings, time reduction, and quality improvements [19].

To represent a real system in a digital twin, it is necessary to combine all available information in a single framework. In addition to sensor data or simulation models, this also includes data on operating settings as well as inspection and maintenance information. Information can be used to show the relationships between the components of a digital twin. This data must be ingested into the digital twin, stored, and transformed so that the algorithms can use it [20]. A challenge remains to combine the data from different sources, across the different interfaces and data formats in real time. No legal or industry standard is yet known for this use case [21].

To ensure continuous and automated data transfer to the digital twin, a suitable infrastructure must be in place. This infrastructure includes, for example, a high-performance, secure Internet connection/speed with which the transmission and processing of information are possible in real-time [22]. In addition, the costs for the required hardware and software should always be compared to the benefits. Setting up the infrastructure can take a considerable amount of time until the various systems are aligned with each other. Likewise, a digital twin should be adaptable to updates and major enhancements [23].

Changes to conditions or new data from the sensors should be able to be adjusted directly in the digital twin without much effort. If individual components or systems are replaced, these changes should also be stored directly in the digital twin. The digital twin is always unique and relates to exactly one object in the real world and collects data over the entire lifecycle [24]. In this way, the digital twin can also receive information from other twins of an object. If an error occurs in the process, the causes of the problem and solutions can be determined more quickly on another product [25].

Artificial intelligence (AI) methods represent another technology for learning a digital twin. These methods can be used to recognize characteristic behavior patterns based on operating data and to identify possible disturbances. The different physical and digital models must be connected in such a way that a constant state can be realized. The different stakeholders require different data or information. In our case, this is related primarily to the recognition of specific objects and their features [17].

Within a digital twin, the data must be prepared accordingly so that each stakeholder receives the information they need on their user interface to perform appropriate actions. Since the digital twin contains all current and historical data of a physical system, it must be secured against unwanted access. Furthermore, the various user profiles must only have access to the information that the profile is allowed to see or receive.

The requirements and prerequisites for creating a digital twin have now been described in detail. Requirements, specifically for the digital twin of a process plant in operation, can also be found in the literature. In their publication, Sierla et al. identify the following requirements [26]:

1. The digital twin precisely records the aspects of the plant that are relevant for retrofitting.
2. The digital twin can be generated from the information of the source, which is usually available at an industrial plant.
3. A minimal manual effort should be required when setting up a digital twin.

In addition, to the requirements from the literature, requirements from the survey of experts can also be listed here. These requirements for the updated digital twin of a process plant can be summarized as follows:

- Capturing the relevant information that the customer needs
- Providing use case-specific information about the actual state of the plant, e.g. Eppinger et al. [27]
- Generation of a listing of the obstructed objects
- Generation of spatial information about solid bodies
- Containing corresponding metadata (e.g., geometric information, administrative information, material information, etc.).

Above mentioned requirements as well as the technical requirements presented in Chap. 2 are essential for a sustainable concept for a 3D digital twin. The developed concept is presented in Chap. 6 in a detailed manner. All requirements will be tailored specifically to the project. At this stage, it is important to understand the requirements for the digital twin in order to gain a better overview.

5.6 Customers' Demand for a Digital Twin

Based on the introduction and description of the customer's view of the digital twin, this section explains the demand in more detail in order to derive and quantify the benefits expected by customers. The need for a digital twin for assets in the process industry is elaborated based on concrete, specific use cases from the industry [28]. It is important to explain the needs at this point since this approach or the conceptual design arose from them [29].

The use of a digital twin offers several advantages that justify its use in a digital environment of an industrial company. For example, as described in the previous chapters, the digital twin enables consistent documentation of a plant's processes over its entire lifecycle. In this way, a better statement can be made about the modernization, optimization, or replanning of a plant. The integration of historical, current, and planned data can also enable better maintenance [30]. Subsequently, the maintenance of a plant can also be simplified, as inspection or maintenance schedules are available. Expected cost savings outweigh the cost of creating and building a digital twin.

A digital twin can support customers in various use cases. As a result, requirements can be derived which help to achieve customer requirements. The needs of a digital twin are described below from the basis and origin of the conceptual design in this approach. After this needs analysis, the decision was made to launch the project of building a full or incremental digital twin of a process plant. Basically, four use cases that benefit from a digital twin have been discovered: (1) as-is documentation, (2) virtual reality, (3) modernization, and (4) augmented reality. These use cases represent the multitude of possible fields of applications for digital twins in the plant industry and can be combined further to create new use cases. They are described subsequently in more detail.

5.6.1 As-Built Documentation

Plants undergo many changes throughout their lifecycle (Fig. 5.5). The first deviations between the planned (as-designed) and the built plant (as-built) usually occur during the construction phase. Although these deviations are known, the models are not updated accordingly, since they do not influence the actual operation of a plant. In terms of documentation, these deviations nevertheless create a problem that needs to be resolved by manual work at each occurrence. Continuous maintenance, repair, and modernization measures result in a constant change in the condition of the plant [31]. The documentation, in particular its update, often lacks in terms of being updated. Main contraction is that there is no solution, which enables an easy update of the digital representation. To ensure an update of a twin on a regular basis a solution is needed, that digitalizes a plant and the undergoing changes on an ongoing basis in an economical manner. This demands a method that works like the automatic update. Such a methodology can be applied as often as necessary. This is the basis to

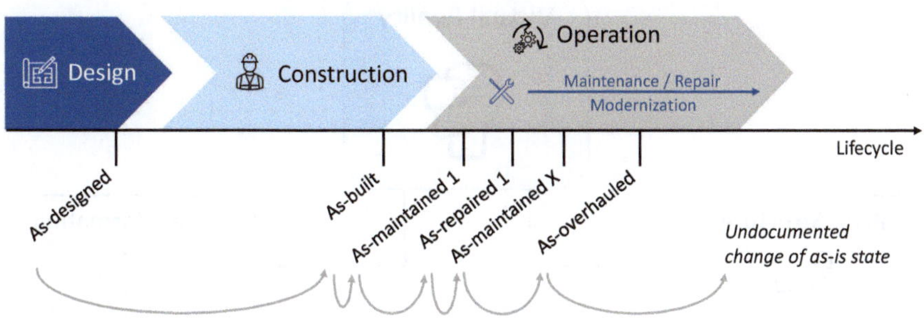

Fig. 5.5 Lifecycle of a plant

ensure an up-to-date 3D digital representation of a plant, which further enables the following use cases.

5.6.2 *Virtual Reality*

During the last decade, the use of virtual representation in virtual reality arose and its exploitation for industrial purposes came into focus. Prominent examples are virtual training and virtual operation. Field engineers or operating engineers can this way be trained on specific plants before they work at that plant in reality. Virtual training is applied when access to a specific plant is not easy such as offshore plants. In addition, if engineers are trained before getting access to a plant the risk of accidents is reduced. A prerequisite for virtual training is up-to-date CAD models which are representing the actual state of the according plant. By using an avatar, engineers can walk through the plants and get familiar with them [32].

Actual models are required with regard to virtual operation as well. The engineer again can walk through the plant with his or her avatar and control instruments like valves or check parameters such as pressure. If the procedure demands closing a valve, the engineer can virtually walk through the plant to the targeted valve and control it virtually. The virtual valve is linked to a physical one in the real plant. The control commands are processed electronically so that the physical valve acts accordingly to the electronic command [33].

The models can be linked with P&IDs and further enriched with sensor data of the plant (Fig. 5.6). For already existing plants, the not-updated and therefore obsolete 3D models can no longer be used due to maintenance, repairs, or modernization activities. The manual creation of a digital twin would require a great deal of effort and would

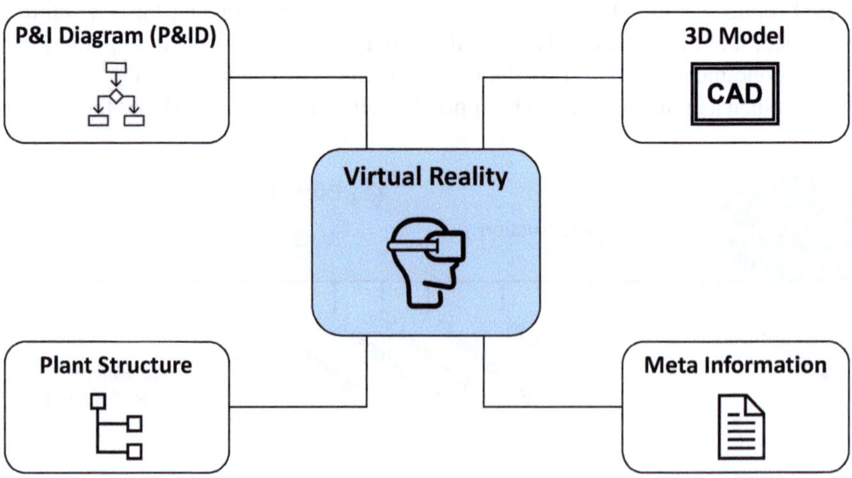

Fig. 5.6 Example image of a virtual reality

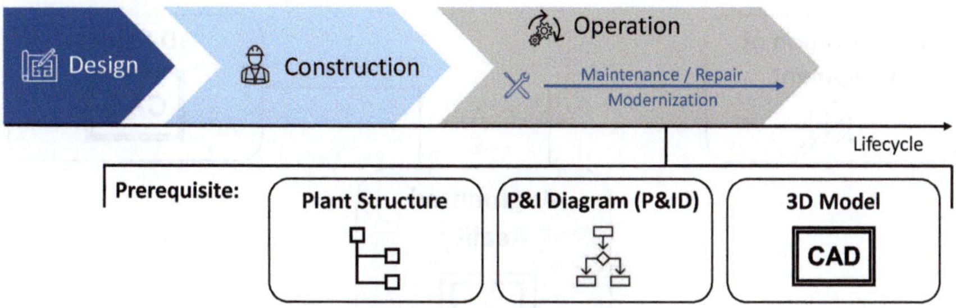

Fig. 5.7 Example data of the modernization of a plant

therefore be very cost-intensive. This is the field for an automated generation of a digital twin.

5.6.3 Modernization

As-is documentation or description of the actual situation in terms of a 3D model is required at the latest when a plant needs to be modernized. These models are the basis for the planning activities for modernization. In many cases, only parts of a plant are renewed and others are kept. It needs to be ensured, that the newly designed parts do fit the remaining parts of the plant. Nowadays, therefore 3D models are remodeled manually to derive a digital 3D representation of the actual plant. This leads to a very high-cost share which could be reduced by an automated generation of a digital twin (Fig. 5.7).

5.6.4 Augmented Reality

To enable augmented reality e.g. for the support of maintenance work, a variety of information is needed. As presented in Fig. 5.8, augmented reality provides benefits in case different sources of information are brought together. The components of a plant must be completely known and clearly identified. The 3D models support the field engineer in quickly understanding the structure of a component. This is the basis that can be enriched with metadata. The linkage between a 3D model of a component linked to metainformation, e.g. from data sheets and structural information, provides the context a field engineer needs to carry out sufficient maintenance. At the same time changes, e.g. replacing a part, can be documented through a mobile device.

Nowadays the documentation of maintenance activities is done manually on hard-copies. Replacements of parts are documented in the printed P&IDs not to mention keeping the according CAD files up-to-date. A digital twin offers the basis for

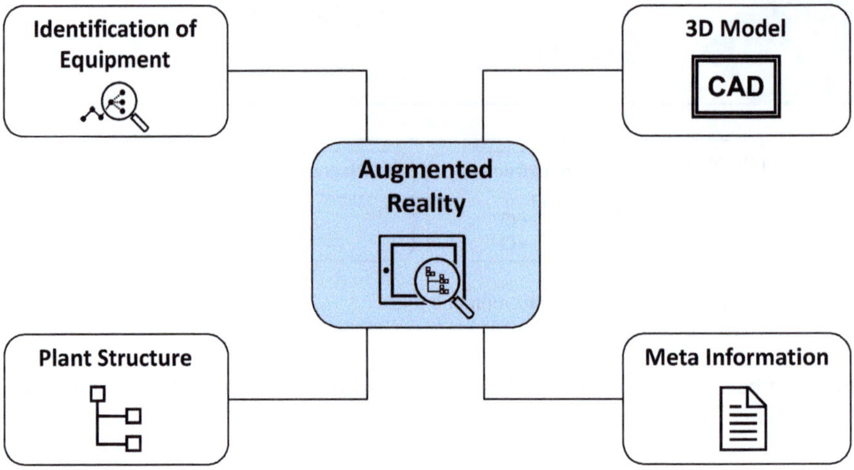

Fig. 5.8 Example for the use in an augmented reality

applying augmented reality for various use cases. It covers all relevant information and sets them into context so that they can be provided for applying augmented reality.

Training with immersive technologies provides an opportunity to incorporate dangerous emergency situations into the training that could not be performed in real-life classroom settings. The possibility of training without being exposed to these risks and practicing tasks safely (in the virtual environment which in the real world would be too dangerous or not possible to perform, and very expensive to organize or reproduce), led to immersive experiences gaining popularity as employee training [33].

5.7 Conclusions and Outlook

The concept of digital twins provides various benefits for different stakeholders in the process industry. Especially when a 3D representation of the actual state is considered. Nowadays, it is necessary to remodel the assets manually, which is time-consuming and expensive respectively. For that reason, 3D representations are only considered for use cases that provide more cost-saving in comparison to the costs spent for the remodeling of the assets. The benefits must be clearly expressed in a corresponding business model [34].

Even though various use cases can benefit from 3D digital twins, the generation is often too expensive, and 3D digital twins are not available for them. Further, 3D digital twins demand a continuous update to represent the as-is state of a plant. Therefore, it is a repeatable task that may include different units in a process plant and be conducted incrementally on demand. An automated generation of a digital twin with spatial information enables stakeholders to exploit digital twins also for additional use cases such as maintenance, modernization, virtual and augmented reality. Further applications are conceivable and can be derived from the current use cases [35].

This envisaged service has to provide clear benefits for the customers. For a successful solution, the requirements from the stakeholders need to be translated into technical concepts (Chap. 6) and implemented (Chap. 7) accordingly. For the definition of a suitable business model, it needs to be elaborated on how such a solution is positioned in the market and competitors need to be identified. The method of buyers experience matrix identified that a solution offering an automated generation of a digital twin with spatial information can be dedicated to the area of the blue ocean [15]. A comparable commercial solution or a direct competitor is not known. The solution contributes to the buyer's experiences in a way that the solution provides major benefits to the customer which easily can be quantified [36].

By applying the method of business canvas the business model has been shaped and subsequently refined. The value proposition has been defined and the customer profiles within a dedicated group identified. Besides the channels to the customer and the customer relationship also the revenue stream and the cost structure has been defined. Additionally, key partners, key roles, and key resources have been identified to ensure a successful solution offering.

The derived business model reflects the current state presented by the existing ecosystem [18], market observation, and customer interviews. Through feedback from experts, the solution will evolve, and the offering might be expanded or adapted. Further, an expansion to other branches can add other customer segments and might require the offering of other order models and revenue streams. In such cases, the business model needs to be adapted or updated accordingly based on the approach presented here.

References

1. Drucker PF (2006) Innovation and entrepreneurship: practice and principles. Harper Business, New York
2. Higgins ET, Scholer AA (2009) Engaging the consumer: the science and art of the value creation process. J Consum Psychol 19:100–114. https://doi.org/10.1016/j.jcps.2009.02.002
3. Adams C (2017) Conceptualising the contemporary corporate value creation processes. Account Auditing Accountability J 30(4):906–931. https://doi.org/10.1108/aaaj-04-2016-2529
4. Mignon I, Bankel A (2022) Sustainable business models and innovation strategies to realize them: a review of 87 empirical cases. Bus Strategy Environ 1–16.https://doi.org/10.1002/bse.3192
5. Freudenreich B, Lüdeke-Freund F, Schaltegger S (2020) A stakeholder theory perspective on business models: value creation for sustainability. J Bus Ethics 166:3–18.https://doi.org/10.1007/s10551-019-04112-z
6. Lehoux P, Silva HP, Denis JL, Miller FA, Pozelli Sabio R, Mendell M (2021) Moving toward responsible value creation: business model challenges faced by organizations producing responsible health innovations. J Prod Innov Manag 38:548–573. https://doi.org/10.1111/jpim.12596

7. Lüdeke-Freund F, Rauter R, Gjerdrum Pedersen ER, Nielsen C (2020) Sustainable value creation through business models: the what, the who and the how. J Bus Models 8(3):62–90
8. Bocken NMP, Geradts THJ (2020) Barriers and drivers to sustainable business model innovation: organization design and dynamic capabilities. Long Range Plan 53(4):101950. https://doi.org/10.1016/j.lrp.2019.101950
9. Sánchez-Gutiérrez J, Cabanelas P, Lampón JF, González-Alvarado TE (2019) The impact on competitiveness of customer value creation through relationship capabilities and marketing innovation. J Bus Ind Mark 34(3):618–627. https://doi.org/10.1108/JBIM-03-2017-0081
10. Zhang R, Wang K (2019) Service supply chain research: a conceptual model based on business processes. J Ind Integr Manage 04(03):1950007. https://doi.org/10.1142/S2424862219500076
11. Sjödin D, Parida V, Kohtamäki M, Wincent J (2020) An agile co-creation process for digital servitization: a micro-service innovation approach. J Bus Res 112:478–491. https://doi.org/10.1016/j.jbusres.2020.01.009
12. Hein A, Weking J, Schreieck M, Wiesche M, Böhm M, Krcmar H (2019) Value co-creation practices in business-to-business platform ecosystems. Electron Mark 29:503–518. https://doi.org/10.1007/s12525-019-00337-y
13. Manser Payne EH, Dahl AJ, Peltier J (2021) Digital servitization value co-creation framework for AI services: a research agenda for digital transformation in financial service ecosystems. J Res Interact Mark 15(2):200–222. https://doi.org/10.1108/JRIM-12-2020-0252
14. Hein A, Schreieck M, Riasanow T, Soto Setzke D, Wiesche M, Böhm M, Krcmar H (2020) Digital platform ecosystems. Electron Markets 30:87–98. https://doi.org/10.1007/s12525-019-00377-4
15. Kim WC, Mauborgne R (2015) Blue ocean strategy—how to create uncontested market space and make the competition irrelevant. Harvard Business School Press, Boston
16. Osterwalder A, Pigneur Y (2010) Business model generation: a handbook for visionaries, game changers, and challengers (Strategyzer). Wiley, Hoboken
17. Stjepandic J, Sommer M, Denkena B (2022) DigiTwin: an approach for production process optimization in a built environment. Springer International Publishing, Switzerland. https://doi.org/10.1007/978-3-030-77539-1
18. NN (2023) https://3digitaltwin.opendesc.com/. Accessed 15 Jan 2023
19. Grau M, Korol W, Lützenberger J, Stjepandić J (2021) Automated generation of a digital twin of a process plant by using 3D scan and artificial intelligence. Adv Transdisciplinary Eng 16:93–102. https://doi.org/10.3233/ATDE210087
20. Sierla S, Azangoo M, Rainio K, Papakonstantinou N, Fay A, Honkamaa P, Vyatkin V (2022) Roadmap to semi-automatic generation of digital twins for brownfield process plants. J Ind Inf Integr 27:100282. https://doi.org/10.1016/j.jii.2021.100282
21. Sharma A, Kosasih E, Zhang J, Brintrup A, Calinescu A (2022) Digital twins: state of the art theory and practice, challenges, and open research questions. J Ind Inf Integr 30:100383. https://doi.org/10.1016/j.jii.2022.100383
22. Stjepandić J, Sommer M, Stobrawa S (2022) The commercialization of digital twin by an extension of a business ecosystem. In: Stjepandić J et al (eds) DigiTwin: an approach for production process optimization in a built environment. Springer Nature, pp 205–233. https://doi.org/10.1007/978-3-030-77539-1_10
23. Kremer P, Lützenberger J, Müller F, Stjepandić J (2022) An approach for the incremental update of a digital twin of a process plant. Adv Transdisciplinary Eng 28:310–319. https://doi.org/10.3233/ATDE220660
24. Gao L, Jia M, Liu D (2022) Process digital twin and its application in petrochemical industry. J Softw Eng Appl 15:308–324. https://doi.org/10.4236/jsea.2022.158018
25. Sanchez-Londono D, Barbieri G, Fumagalli L (2022) Smart retrofitting in maintenance: a systematic literature review. J Intell Manuf. https://doi.org/10.1007/s10845-022-02002-2
26. Sierla S, Sorsamäki L, Azangoo M, Villberg A, Hytönen E, Vyatkin V (2020) Towards semi-automatic generation of a steady state digital twin of a brownfield process plant. Appl Sci 10(19):6959. https://doi.org/10.3390/app10196959

27. Eppinger T, Longwell G, Mas P, Goodheart K, Badiali U, Aglave R (2021) Increase food production efficiency using the executable digital twin (XDT). Chem Eng Trans 87:37–42. https://doi.org/10.3303/CET2187007
28. Qi Q, Tao F, Hu T, Anwer N, Liu A, Wei Y, Wang L, Neef AYC (2021) Enabling technologies and tools for digital twin. J Manuf Syst 58(Part B):3–21. https://doi.org/10.1016/j.jmsy.2019.10.001
29. Perno M, Hvam L (2020) Developing a framework for scoping digital twins in the process manufacturing industry. Adv Transdisciplinary Eng 13:475–486. https://doi.org/10.3233/ATDE200185
30. Bamberg A, Urbas L, Bröcker S, Bortz M, Kockmann N (2021) The digital twin—your ingenious companion for process engineering and smart production. Chem Eng Technol 44:954–961. https://doi.org/10.1002/ceat.202000562
31. Sutton I (2015) Plant design and operations. Elsevier, Waltham
32. Eswaran M, Raju Bahubalendruni MVA (2022) Challenges and opportunities on AR/VR technologies for manufacturing systems in the context of industry 4.0: a state of the art review. J Manuf Syst 65:260–278. https://doi.org/10.1016/j.jmsy.2022.09.016
33. Garcia Fracaro S, Glassey J, Bernaerts K, Wilk M (2022) Immersive technologies for the training of operators in the process industry: a systematic literature review. Comput Chem Eng 160:107691. https://doi.org/10.1016/j.compchemeng.2022.107691
34. Rupčić N, Majić T, Stjepandić J (2020) Emergence of business ecosystems by transformation of platforms through the process of organizational learning. J Ind Integr Manage 5(2):181–203. https://doi.org/10.1142/S2424862220500086
35. Stecken J, Ebel M, Bartelt M, Poeppelbuss J, Kuhlenkötter B (2019) Digital shadow platform as an innovative business model. Procedia CIRP 83:204–209. https://doi.org/10.1016/j.procir.2019.02.130
36. Mihăescu L (2021) Economic intelligence: using innovation to reinvent the business. In: Orăștean R, Ogrean C, Mărginean SC (eds) Organizations and performance in a complex world. IECS 2019. Springer Proceedings in Business and Economics. Springer, Cham. https://doi.org/10.1007/978-3-030-50676-6_14

Chapter 6
Solution Approach for Digital Twin of a Process Plant

Abstract This chapter presents how to conceptualize the solution approach for a digital twin of a process plant, after the requirements, the state-of-the-art, and business needs are explored. The servitization process is a fundamental means for manufacturing companies that would find new business opportunities and involve new customer segments, increasing their market share. In a working environment where only a few standard procedures exist, ideation gets special importance. Based on a solution search, the product-service development is conducted from the initial requirements to the final customer service. For this purpose, a productive solution is re-used and adapted for a new industry with specific customer requirements. Although the procedure starts again with a scan of the process plant and ends with an editable CAD model for the re-design of objects in question, it has fundamental changes in its structure and singular steps and can be considered as an entirely new solution within a platform. Based on the process engineering introduction to pipeline construction, an object tree that maps the basis for the concept is created. The solution is drawn as a process chain with several steps which work like a 3D search of the components of the digital twin in the point cloud. A four-step methodology for the semi-automatic generation of digital twins from the point cloud of the plant is explored. The Zachman framework is then applied according to an already-used approach which determines the functional description of the legacy system. Different actors (customer, capture partner, cooperation partner) in designed service during the emergence of a digital twin are described, in particular their roles, and influence on the customer, and how they supply corresponding data and information. The object recognition process and the impact factors on the immediate and expected final results are explained. Translation of the piping system description from the neutral to the proprietary environment (e.g., AVEVA) is emphasized with relevance to the following engineering design which is briefly explained on the AVEVA's E3D mode of operation. Additional complexity is given in this approach by collaborative execution with an engineering partner. Therefore, the work share with this partner is described in detail.

J. Stjepandić et al., *Generation and Update of a Digital Twin in a Process Plant*,
https://doi.org/10.1007/978-3-031-47316-6_6

Keywords Digital Twin · Business value · Ideation · Machine learning · Engineering collaboration · Object recognition · Piping and instrumentation diagram

6.1 Introduction

While a product represents the tangible commodity manufactured to be sold and capable of fulfilling the users' needs, a service is understood as the "activity" delivered to generate an economic value by its exploitation and often done on a commercial basis. Indeed, the servitization process is a fundamental means for manufacturing companies that would find new business opportunities and involve new customer segments, increasing their market share. This process not only affects the company business model but also the whole enterprise, in terms of those internal processes and standard procedures that support the design, development, and delivery of the new value proposition [1]. Service innovations challenge existing offerings and business models (as presented in Chap. 5), shape existing markets, and create new ones, in particular by creating digital platforms [2]. But every product is also a service—seen as the lifetime of value it delivers every time being used. One way of looking at this service model is that the service is the sum of the user's experiences with the product, and incorporates everything from how the product is initiated, used, consumed, updated, and even discarded [3]. When conceptualizing a new idea for a solution, it is crucial to identify the most important dimensions and find answers to specific questions in order to propagate the idea from the initial thought through the various stages of innovation [4].

Initially, a comprehensive study that encompasses conceptual investigation and qualitative data, gained in in-depth interviews with stakeholders (as presented in Sects. 2.2 and 5.7), is necessary to conceptualize "smart service" and "smart service systems" based on using smart products as boundary objects that collect service consumers' and service providers' resources and activities. Smart services facilitate both actors to get and analyze aggregated field evidence and to adapt service systems based on contextual data. Subsequently, boundary objects enable dual value creation. On the one hand, the user benefits from the frontstage usage of the product by creating and capturing value-in-use (by monitoring, through optimization, or remote control). On the other hand, the provider benefits from backstage analytics, such as remote monitoring and diagnostics, data aggregation, data analytics, or decision-making [5].

While research on the business value of IT initially focused on the internal company-level perspective, applications in the era of digitalization have moved the focus to how IT can function as an integral part of products and services, which yields new digital business models. For this purpose, 7 dimensions of a framework are elaborated: initiation; development; implementation; exploitation; the role of the external competitive environment; the role of the internal organizational environment; as well as product, service, and process outcomes [6]. The value can be generated in four strategic categories: improving returns on assets, re-engaging customers, creating new products and business lines, and transforming business models [7]. Therefore, the impact of IT shifts the company-level perspective to a value chain perspective. Conversely, the digital twin creates value in internal processes and/or in products and services. Subsequently, the digital twin generates value for users and providers alike, referred to as frontstage value and backstage value, respectively. The business value of IT and digital twin is compared in Table 6.1 [8].

Ideation is the process of searching for a solution to a particular design issue. When classifying a situation that is the subject of ideation methods, the possibilities are basically inexhaustible. Especially in complex systems, the characteristics are usually very different from each other. This necessitates, on the one hand, considering the solutions individually, and, on the other hand, great difficulties in developing a procedure for determining the most appropriate method to solve them. The second major obstacle is the difficulty of formulating a complete classification of idea generation methods for a dedicated solution [4].

The challenge of classifying ideation methods is also related to the difficulty of verifying multiple aspects. An example would be the structuring of idea-generation techniques into single methods and group methods, as we find them in our case. This structuring does not cause any major problems in the procedural sense. In most cases, a procedure of a particular method determines the assignment to one of two categories. If the classification is not aligned with formal aspects, such as the quality of ideas and the mechanisms of their generation, depending on the number of people

Table 6.1 Comparison of the business value of IT and digital twin, derived from [8]

	Business value of IT	Business value of digital twin
What is the value perspective?	Mostly internal, company-level perspective	Value chain perspective covering multiple internal and external stakeholders in B2B markets
Where is the value created?	Predominantly in internal processes; indirect effects on market performance considered	In internal processes and/or in products and services
How is the value created?	Predominantly by enhancement of internal processes; recent perspectives also consider IT as part of new offers	Enhancement of existing and creation of new processes and/or products and services that may lead to performance improvements
Who benefits from value creation?	The IT solution user benefits directly from single value creation through value-in-use (*frontstage value*)	The digital twin user and the digital twin provider benefit from dual value creation enabled by the digital twin, i.e., the user benefits directly (*frontstage value*), and the provider benefits indirectly (*backstage value*)
How can the value be increased?	Predominantly by streamlining internal processes	Extension of internal processes to the value chain; transforming the business model

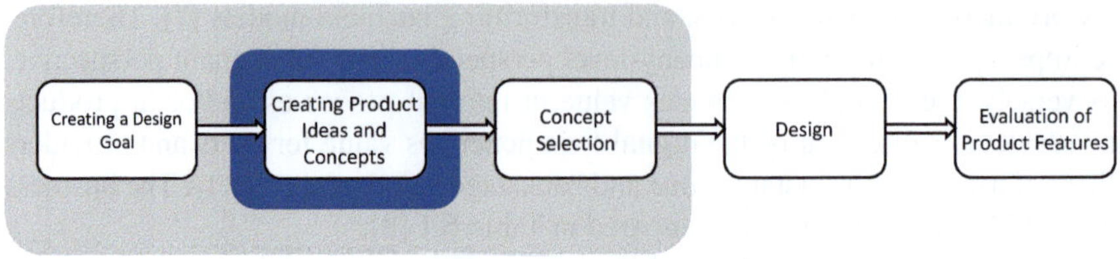

Fig. 6.1 Steps of the design method, derived from [4]

involved in the ideation process of a particular method, the situation becomes more difficult. A lack of data on this topic, as well as difficulties in collecting them, makes it impossible to structure idea generation techniques in a way that would be meaningful from the standpoint of method choice, elaboration of organizational assumptions, and organization of the design team [4].

In the narrow sense, ideation is understood to find as many solutions as possible for the concrete issue [9]. The selection and evaluation phase is primarily considered an integral part of ideation, but it can be considered as an independent, subsequent process based on the results of ideation [10]. Then ideation is understood as only a procedure for the generation of a set of solutions that are expected with a high probability to be the right ones. In many cases, the decision on the selection of the concept does not occur directly but is done in an indirect way during the implementation of the algorithm of a specific ideation method [11].

Figure 6.1 discovers two approaches to the ideation process:

- A more or less formalized final concept,
- A set of possible variants of the solution.

Firstly, the generation of one specific concept, which is potentially a solution to the initial problem. Secondly, the ideation stage in the design process is a certain set of potential solutions. These solutions are analyzed further to enable decision-making about the choice of a suitable concept, frequently using criteria. The presented division may seem a little artificial, but it answers the important question, namely, at which point the ideation process ends and when the optimization process of the concept begins [4].

In this work, the product-service development is conducted from the initial requirements to the final customer service [12]. For this purpose, a productive solution is re-used and adapted for a new industry with specific customer requirements [13]. Although the procedure starts again with a scan and ends with an editable CAD model for the re-design of the objects in question, it has fundamental changes in its structure and singular steps and can be considered an entirely new solution. Additional complexity is given by the collaborative execution of this procedure with an engineering partner [14].

The remainder of this chapter is structured as follows: in Sect. 6.2, the methodology for concept selection is presented with the basic solution approach and solution blocks. In Sect. 6.3, functional components of the piping system are explored based on the process engineering point of view. A description of the actors and the collaboration among them is presented in Sect. 6.4. Finally, Sect. 6.5 summarizes the conclusions and outlook.

6.2 Concept Selection

Based on the customers' requirements and the business case, a solution approach for the incremental updating of a digital twin in a process plant is derived and explained in this chapter [14]. Based on the process engineering introduction to pipeline construction, the first step in this chapter is to create an object tree that maps the basis for the concept. This object tree is adapted to the "IT process view" so that it is made clear which objects the digital twin can contain.

Subsequently, the stakeholders of the overall system are identified with the help of the Zachman Framework, which is used to translate business requirements into technical solutions [15]. For this purpose, there is a short introduction before the application of the Zachman Framework is executed accordingly, as shown in previous work [13]. This application forms the foundation for gaining an overview of the various elements that are related to each other in the creation of the digital twin. Finally, the individual elements of the overall system are presented and explained in more detail.

Once the elements of the overall system have been identified, requirements are collected for the digital twin, which emerges at the end of this conceptualization. Based on the requirements, the processes are then elaborated on and agreed upon before they are verified. The modeling of the processes and the illustrations are done with the software "Cameo Systems Modeler" (see Sect. 2.2.1).

6.2.1 Solution Patterns

Basically, the digital twin is built through an interpretation of the point cloud, supposedly of sufficient data quality [16], acquired by an appropriate object acquisition device (e.g., scanner) [17]. This procedure is automated by using machine learning (ML) methods and derived object recognition. The pre-requisite is that the ML has been trained appropriately beforehand with a sufficient volume of data [18]. At the end of this procedure, it may be necessary that the point cloud of the plant or the 3D models derived from it are linked with additional information from the piping and instrumentation diagram (P&ID). This helps to identify the plant components and their function reliably (Table 6.2, Fig. 6.2). This can be provided in collaboration of two partners.

This idea should be refined further by creating a set of appropriate functions that process and transform the point cloud to the desired final results. For this purpose, a SILCACC toolbox was composed of 7 functions that define the actions: Segment,

Table 6.2 A SILCACC set of functions to create a digital twin from a point cloud

Action	Aim	Input	Output	Method
Segment	Subdivide a point cloud into multiple portions of interest according to various criteria	Point cloud	Point cloud (segments)	AI (ML)
Identify	Determination of an object or an object cluster in a point cloud by comparison with pre-trained objects ("3D search")	Point cloud	Pointer to a sample object in the database	AI (ML)
Locate	Determination of the spatial position and orientation of an object	Point cloud (cluster)	Translation matrix	Deterministic
Check	Verification and validation of desired results	Any	True or false	Depends on the object type
Adjust	Adapt the parameters of an object to the constraints (e.g., adjacent objects)	Dimension, Constraint	A slightly changed object	AI (ML) and/or deterministic
Concatenate	Generate a chain of adjacent objects	A set of objects	3D structure	AI (ML) or deterministic
Connect	Create a link to external models and data (e.g. piping and instrumentation diagram) for the aim of integration or communication	A set of objects	Linked objects and/or data (2D/3D)	AI (ML) or deterministic

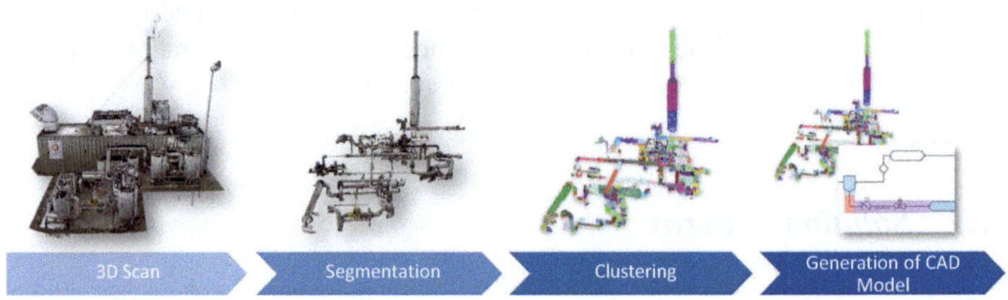

Fig. 6.2 Generalized process of object recognition and generation of CAD models

Identify, Locate, Check, Adjust, Concatenate, and Connect (Table 6.2). This set of functions should be enough to fully cover the process from the point cloud to the digital twin. Of course, each function can be subdivided and refined further. These logical functions should fulfill the overall criteria for the architecture of components of a complex software system [19] and be automated accordingly. Particular attention is paid to the piping systems due to their quantity and importance in case of design change [20].

By generating the 3D scan of the existing plant or parts of it (e.g., in case of an update), the current status of a plant is captured and represented as a point cloud.

Standard commercial devices (terrestrial, mobile, carried by a drone) can be used solely or combined for this purpose, depending on the extent and complexity of the plant. For the caption of details, a consumer camera can be used too.

While scanners usually deliver point clouds of a higher density than necessary for reliable object recognition, the first process step comprises the reduction of points to a necessary abstraction level for object recognition [17]. This significantly reduces the amount of data and makes it manageable.

6.2.2 Solution Blocks

Since the designer is primarily interested in the structure of the piping system—described by the course of the center line—as well as the attributes and the position of individual components (fittings, equipment)—the solution must consist of individual function blocks that quickly lead to this goal and provide interim results that facilitate check of results. While the amount of data between a scan and the desired final description must be reduced from terrabytes to megabytes, the individual function blocks must also perform a certain filter function [21].

Therefore, the solution consists of a process chain with several steps which work like a 3D search in the point cloud. First of all, the components of the piping system must be extracted for a local search. This block works like a global segmentation (see Table 6.2; A: Segment). Further constituents of a plant (equipment, piping, structure) can be extracted as well. The results are the partial point clouds that comprise objects of the dedicated category [22].

In the next block, the 3D search is performed in the partial point clouds. Singular clusters must be extracted which comprise one part (A: Identify). This task is conducted by a pre-trained convolutional neural network (CNN) which promises the best results at this time. Subsequently, the component recognition runs in the singular clusters. It also includes the identification of the main or determinant parameters (e.g., length, diameter of a piping segment). This task is also performed by appropriate pre-trained CNN [23, 24].

When all the components are recognized and exactly localized (A: Locate), the recognition of piping system segments should start with the objective to rebuild the piping system from scratch [25, 26]. This search should primarily use the appropriate geometric characteristics such as coaxiality, a consistent centerline, etc. (A: Concatenate).

After the piping system is rebuilt, the linking with P&ID from a commercial CAD system for plant design (e.g., E3D from AVEVA) can be performed (A: Connect). This functional block must comprise a fallback approach if gaps occur in a piping system segment (A: Adjust) [27].

6.3 Functional Components

Based on the process engineering fundamentals of a piping system, the fundament for the conceptual design is now laid. For this purpose, the piping system is suitably modified for the concept from the process. To illustrate this, the components of the piping system are shown in the following Fig. 6.3.

Figure 6.3 provides an overview of components described in the basics of piping construction (see Sect. 4.3). The overview (Fig. 6.3) is structured according to recognizable objects in a process plant from an IT perspective and not from the process engineering view. All mentioned components can potentially become part of a digital twin, that is to be created as a result at the end of this concept.

Subsequently, the components are described in order to explain the relationships in the object tree. It contains process parameters, which have an essential meaning for the concept. Following the literature of K. Sattler and Kasper [28] the further subdivision into the individual components takes place. They are further divided to provide the most accurate view of a process plant from the IT process perspective. The object structure represents the basis of the concept, it is nevertheless only the first step in the total process. A hierarchical representation from "big to small" until the individual components of a unit are provided by the object tree. A breakdown down to the smallest part of the overall plant is provided.

At the top left of Fig. 6.3, the outline of the piping plant is depicted. The outline is shown in white blocks with a black frame and black text. As mentioned in Sect. 4.3, a process plant is made up of different piping systems. These, in turn, are made up of pipelines. The piping branch or the piping section is part of the pipelines. Up to this point, the piping system has been divided.

Due to the limitation of the CNN framework's recognition capability, only the objects of a piping system that are distinctive enough to recognize them from a point cloud were included. In the figure, it can be seen that a piping branch is composed of physical components such as piping parts, equipment, instruments, storage, and supply units. It is also possible to enrich the digital twin with metadata to optimize it.

Piping parts make up the largest percentage of components in a process plant. Different types of piping parts can occur, such as straight pipes, fittings, connecting elements, or valves. These are shown in the object tree with a red box and black font.

An essential part of piping parts is the type of straight pipes. It does not lead to any change of direction of the flow material in a process. Another large part of the piping parts is the fittings, which are performed by piping parts [29]. Fittings are subdivided into elbows, reducers, tees, intersections, and plugs. There are ways to continue the subdivision further, but from the IT process perspective, these are the most prominent components recognized by the CNN framework. Here, the elbow is an angled piece of pipe in which there is a change in the direction of the flow material. The pipe elbow has the same diameter throughout. A reducer is a straight pipe that has a different inlet and outlet diameter. The tee is a way to insert a branch into the process. It is usually the case that a main pipe from the main process has the

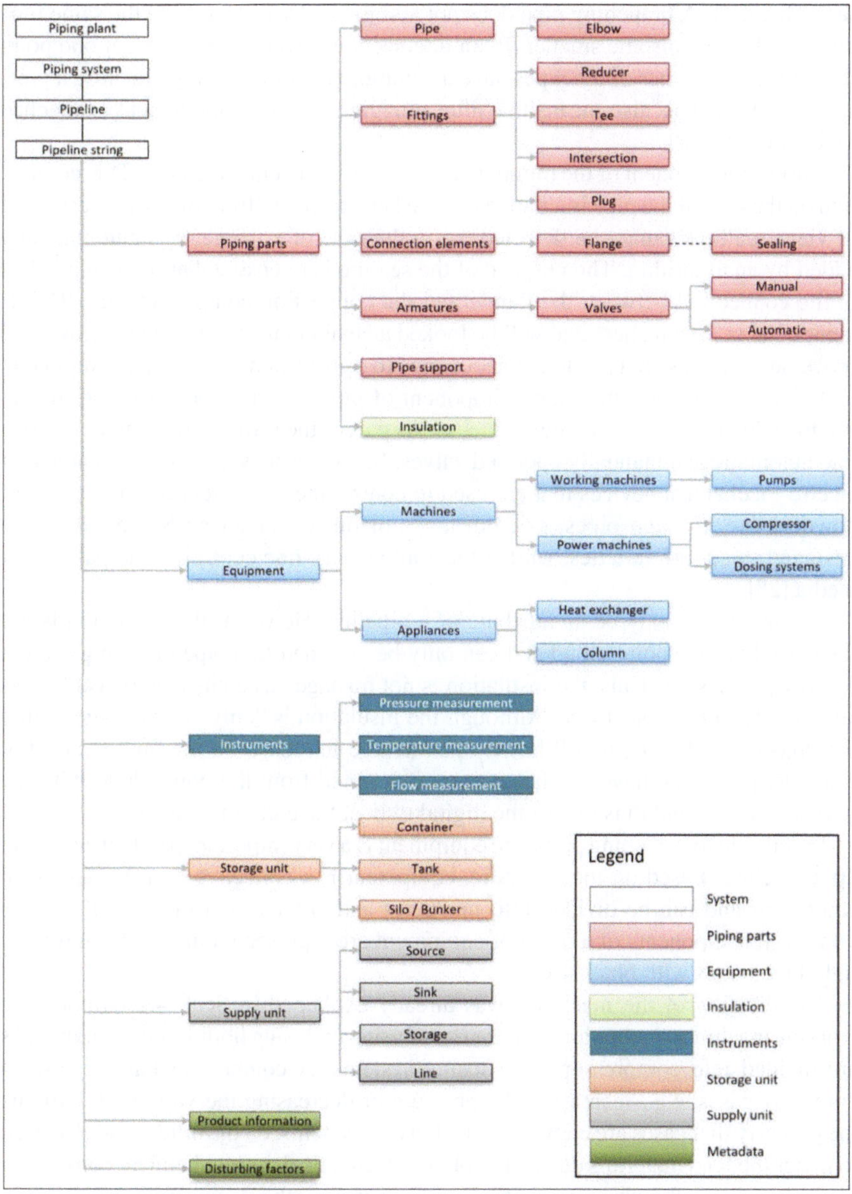

Fig. 6.3 Object tree plant engineering from an IT process perspective

same diameter. A branching pipe does not automatically have to have the same pipe diameter but can also be smaller. In an intersection, several pipes meet at one point so that the centerlines of the pipes have a common point of crossing. To close a pipe, a plug is used. This also means that after plugging, no other component is attached to the process [28].

Another component of the piping parts can be a connection element. This connection on the side of the pipe is a flange or a gasket. However, from the IT process point of view, sealings cannot be detected, so in the case of a flange, it is automatically added by an algorithm. The purpose of the sealing is to ensure that no media is lost in the connection between the flange and the connection component [28]. This is mentioned in passing here and will be looked at again in more detail in the flows and processes. Flanges are used to connect a pipe to a corresponding mating component.

Valves are another important component of piping parts. They are used to shut off, throttle, or regulate the flow of media. As part of the fittings, they can be divided into automatic and manually operated valves. In this sense, valves are mechanical or electro-mechanical devices that are used to control the movement of liquids, gases, powders, etc. through pipes, or from tanks or other containers. Some valves are designed as on–off varieties, while others allow very fine control of the passage of media [28].

The insulation may be an attribute of a pipeline. However, the insulation is not recognizable by a point cloud and can only be assigned to a pipe or piping section by post-processing. Thus, the insulation is not managed as a physical object, but as an attribute in the object tree. Although the insulation is "only" an attribute within the object recognition, it still has a considerable influence on the final result. For example, pipes that have insulation can be detected from the point cloud with the wrong diameter and thus distort the digital twin at the end of the process.

In addition to the piping parts, the equipment is a very important part that can make up a pipeline. Based on the literature, equipment is a collective term for technical equipment and can be divided into machines and appliances from the IT process view. The components of a piece of equipment are represented in the object tree by light blue boxes with black text.

The division of the machines was already explained in Sect. 4.2. Among the working machines, pumps are used for transferring or dosing liquids. Power machines are divided into compressors and dosing systems. A compressor has the task to compress gases, e.g., increasing the pressure or decreasing the volume. It exploits the property that gases are compressible. Dosing systems, on the other hand, are used to meter the feed materials during the physical, chemical, or biochemical conversion of substances in the process plants. For this purpose, the dosing systems differ in that substance quantities are formed, a component is added or a process variable is controlled [28].

Material transformations and heat exchanges take place in the appliances such as heat exchangers or columns, which are the most important examples. The heat exchangers have the task of transferring heat from the emitting to the receiving medium. Columns, on the other hand, are used in a process plant to carry out thermal separation operations such as rectification, distillation, absorption, desorption, and extraction [28].

Instruments (aquamarine boxes with white lettering in Fig. 6.3) are used to measure or control the important process variables. Therefore, instruments are the placeholder for the components of the measured value acquisition and processing. The most important ones have been included in the object tree. Thereby the instruments are directly integrated into the process and are not connected to the process by a flange etc. It is difficult to filter from a point cloud what kind of measurement it is. Nevertheless, this can be clearly defined by appropriate post-processing or attached information. Usually, pressure and temperature are measured in a process plant [28].

Storage units are represented in the object tree by orange boxes with black text. They can store raw materials, intermediate materials, or the final product. As already mentioned in the basics, these can be divided into containers, tanks, or silos/bunkers.

Another important part of a process plant is the supply units. These are made up of sources, sinks, and storage. One of the biggest difficulties in building a digital twin is distinguishing the supply system from the main process from a point cloud.

In order to match a digital twin of a process plant as closely as possible to the physical twin, it is important to have appropriate metadata attached to the digital twin. This metadata is indicated in the object tree by a light green box with white text. Thus, the product information is included in the object tree, which contains a description and an overview of products, operating materials, or raw materials that are involved in a process. Furthermore, the disturbance factors are included under the metadata. This component includes all elements that influence the process but cannot be manipulated. These factors are those that occur externally as well as internally. These are, for example, protective devices, fences, or environmental factors such as trees or clouds.

An overview has now been provided of the parameters that can occur in a process plant from the IT process engineering point of view. The object tree depicted in Fig. 6.3 discovers the multitude of objects in different variations which can be expected in a process plant.

6.4 Collaboration Among Stakeholders

One of the most commonly used methods for a large project that brings a change is the Zachman Framework. This framework is an enterprise architecture (EA) framework, built in a matrix form, usually consisting of six rows and six columns, or 36 cells. It is used to represent and describe an "entity", that can be anything, yet it is usually a company or information system [30].

The description is composed of up to six "views" and six "perspectives", with the rows symbolizing the "view" and the columns symbolizing the "perspective". For each view, there are six perspectives in this approach, which are explained below [30]:

- **Planner's View—Scope:** Can be considered as the summary of a project. Scope summarizes aspects such as overall size, cost, relationships, etc. This is how the scope and context of a project can be defined.
- **Owner's View—Enterprise Model:** It reflects an architect's view. This is from a high level of abstraction of an owner and also includes the business units, such as the processes. These must be fully understood in order to actually use a system.
- **Designer's View—Information System Model:** The owner's view represents a designer's perspective to initiate the implementation. Mostly a technically detailed view, such as data structures, behaviors, partitioning, etc.
- **Developer's View—Technology Model:** An owner's view from the developer's perspective to continue to work with is part of this view. Unlike the designer's view, the focus here is on the implementation tasks and constraints, such as the tools, technologies, materials, etc.
- **Subcontractor's View—Detailed Design:** Includes a detailed design that e.g., a programming team or a builder can work with. Among other things, this design contains complete specifications for required systems or subsystems. In addition, in this view, also all information about configurations, creation, and deployment can be found.
- **View of the System (Current):** Corresponds to the actual system that is to be developed, built, or modeled.

To use the Zachman framework, not all views need to be defined, yet it is common to use all six. However, the six perspectives must be described [30]:

- **What?—Data description:** It includes the most important elements and their relationships to each other.
- **How?—Function description:** Outlines the function and behavior of an element as well as the behavior together.
- **Where?—Network description:** Description where the system elements are used. In addition, this view states all dependencies between the elements.
- **Who?—Description of the person:** Contains the description of which persons are involved in the system.
- **When?—Description of time:** Temporal aspects of the elements are considered here.
- **Why?—Description of motivation:** Describes the requirements and the basic reasons for the system.

The Zachman framework is then applied according to an already-used approach which determines the functional description of a legacy system [31]. This application is important to ensure the coherence of the system. This work builds further based on this application. Subsequently, a package diagram is presented so that an appropriate overview of the elements in the overall process is obtained (Table 6.3).

While the digital twin emerges in a collaboration of several judicial entities with a frequent exchange of data, the most important actors from the overall process (client, capture partner, PROSTEP AG, optionally: cooperation partner) are shown

Table 6.3 Application Zachman Framework [31]

	What	How	Where	Who	When	Why
Scope	Technical equipment provided by a customer or its partners and submitted according to specific specifications	Capture and exchange are offered as a cloud service	Global: All technical locations of the parties involved are affected	Plant operators, system suppliers and planning offices of a customer	Whenever a design change is made by a party involved	Cross-company collaboration without the need for on-site infrastructure
Enterprise model	**Entity**—Companies that place an order and have technical data, moreover, the activities can be tracked by means of a report **Relationship**—Contract (SLA)	**Process**—Change management, onboarding, data transmission, data reception, support	**Knots**—Business locations (mostly technology) **Link**—Connection	**Humans**—Maintenance, design department, service **Work**—Service level agreement, process order	During data transmission, during data reception	Enables technical organizations to realize the need based on fast, simple and reliable communication
(Logical) system model	**Entity**—Electronic records contained in an order and transmitted to a user in the organization with a contractual agreement **Relationship**—Person in the organisation	**Process** = Setup, output/ input data **I/O**—Internet All participants must be registered on the platform in order to use the services	**Knots**—Business unit **Connection**—Web portal, OFTP2	**Humans**—Seller, designer, process planner, administrator, sender, receiver **Work**—Technical design, process planning, sending and receiving data, managing services	On demand/ wish of the customer	A party may be required to provide data in a specific configuration to a customer at a specific time
(Physical) technology model	Model-based physical data models with diagrams of technology structures, control structures, definitions and descriptions are emulated using UML techniques. Users interact directly with the service using (1) a web portal or (2) an application programming interface for machine-to-machine communication. The user interface is described with graphical examples					

(continued)

Table 6.3 (continued)

	What	How	Where	Who	When	Why
Detailed representation	By using templates, each OPENDESC.com client can be easily set up as a node with specific characteristics. Each node is logically separated from the other clients to meet the strict security requirements. The operation of a node is done through an administration and operation interface, which allows updating the node database					
Functioning enterprise	Each authorized user can establish communication relationships, set rules for sharing, timing, quantity, desired data quality, level of detail and intellectual property protection, add required metadata and distribute records to multiple users, etc.					

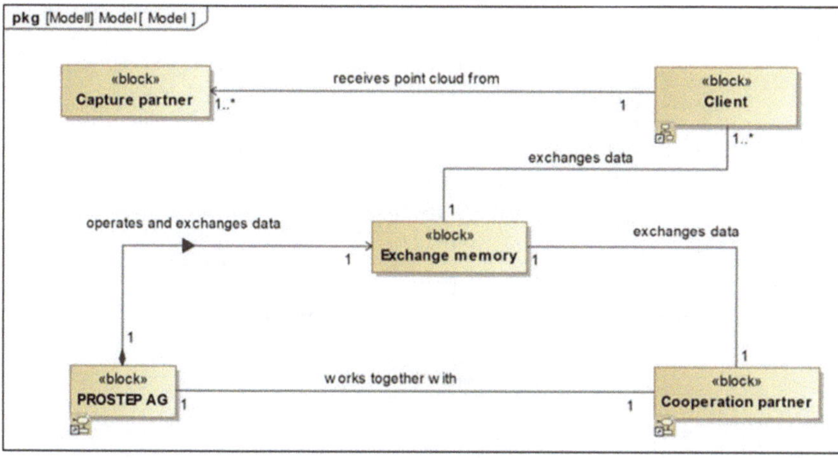

Fig. 6.4 Overview diagram of the elements in the overall process

in a package diagram in Fig. 6.4. This formal process is described based on the Zachman framework [31].

The need for the digital twin originates from a customer. It exchanges data and information with PROSTEP AG and the cooperation partner via an exchange memory. Data is placed in this common repository and the processed data is taken from the exchange memory. Through the exchange of data, the exchange store is the central role in the diagram and the process. The exchange memory is operated by PROSTEP AG. PROSTEP AG and the cooperation partner take the required data from the memory. A sufficiently accurate and consistent point cloud of a process plant is essential for the entire concept. If this point cloud of the plant does not yet exist, a capture partner must be commissioned by the customer to generate this point cloud.

6.4.1 Client

As depicted in the following Fig. 6.4, the focus is placed on the customer as well as its partners who may contribute to the emergence of the digital twin. Based on market analysis, the engineering, procurement, and construction (EPC) contractor (see Sect. 4.2) is primarily seen as the customer who is on duty for the entire plant project including the digital twin.

Figure 6.5 shows the multitude of actors which have an influence on the customer, or which actors can supply corresponding data and information to the customer for the exchange memory. As can be seen in the diagram, not only the data of a customer is transmitted to the exchange memory, but also ancillary information from system

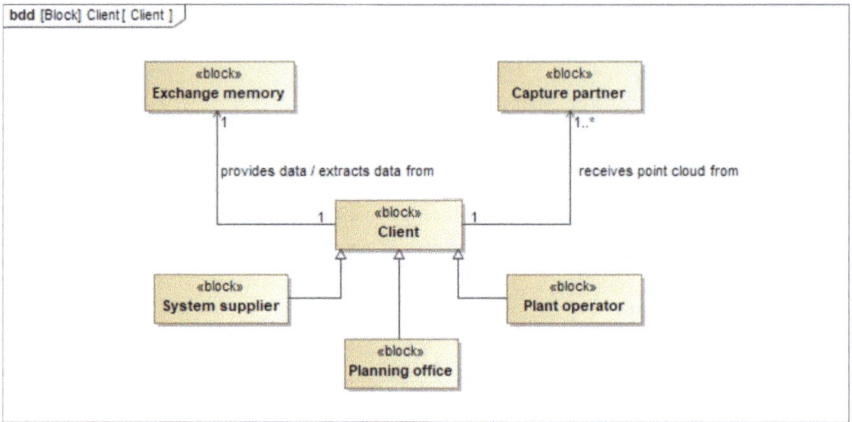

Fig. 6.5 Block diagram client

suppliers, the planning office, or the operator of a plant. This ensures that all the customer's data and information are properly collected for the digital twin.

The process to build a digital twin would not start without a customer. Accordingly, the customer is the start or foundation of any process which is a one-of-a-kind delivery. PROSTEP AG or the cooperation partner is contacted by the client with a need or a request for this digital twin.

As a foundation, a contract exists between a customer and PROSTEP AG in which the terms and conditions are recorded. These are, for example, the scope, the sending, and receiving of data, the timeline, the charges, etc. A contract can also exist between the customer and the cooperation partner. In this case, PROSTEP AG's field of activity would not change, but it would be classified as the role of the sub-contractor.

During the data exchange process, the customer releases the available data. A representative extent of data of a certain quality is required in order to build a digital twin. Therefore, the capture partner has an important role, since he supplies the point cloud data of the plant (Sect. 6.4.1.1), which again will have to be made available to PROSTEP AG. Derived parameters, CAD models, and further can be transferred to cooperation partners via the exchange memory.

6.4.1.1 Capture Partner

The capture partner is usually commissioned by the plant operator regularly for the sake of documentation of the built status which may change periodically. If no point cloud of the process plant is available, it must be generated as the input for the digital twin. The capture partner plays a decisive role in the internal process of PROSTEP. Basically, he has three technical alternatives to acquire a point cloud of the process plant [20]:

- **Laser scan method:** Laser scanners uses laser light to capture the environment. Each time the emitted laser hits a surface, the light is reflected. This reflected light is then picked up by the receiver and deflected by a deflection mirror, which is set in rotation. That is repeated (several hundred thousand times per second). The received back laser light is then automatically analyzed and evaluated.
- **Digital methods based on a 360° camera:** A 360° camera allows images to be captured in all directions, both vertically and horizontally. In other words, the camera is the center of the captured image. This technique requires a high level of computing power. All individually captured images are then stitched together to form an overall image with an all-around view. The light rays are reflected both horizontally and vertically onto a wide spectrum of the lens using a lens and a mirror. This makes it possible to obtain a 360° scan. Pre-condition is that all light rays are projected onto the lens.
- **Acquisition with a mobile device:** When scanning with a mobile device, the environment is recorded by appropriate hardware and software. Such devices can be, for example, a cell phone or a tablet. The corresponding algorithms merge the images in the background and a scan of the environment is created.

Which of the three methods is used to generate the point cloud is irrelevant at first. However, the point cloud must meet certain requirements. These requirements are depicted in more detail in the following Table 6.4.

Table 6.4 Requirements for the scan, derived from [20]

Characteristics	Description
Quality	The quantity and density of the points are quality features. The distance between the two points should be no more than 5 mm.
Registration	As a rule, scans must be performed from several locations in order to capture all angles and areas. All individual scans must be correctly positioned to each other afterward to correctly depict the captured plant. This process of correct alignment is called registration.
Color	The scan must consider color information.
Recording project	Optionally, the exposure project, which consists of all partial exposures, should also be transmitted. This makes it possible to carry out the corresponding optimizations.

6.4.1.2 Cooperation Partner

The digital twin project is carried out based on principles of collaborative project management in collaboration with several partners which contribute to their roles and dedicated work packages. A specific role comprises CAD design within the digital transformation in process plant engineering and construction which heavily depends on the used toolset (e.g., AVEVA, CADMATIC, Hexagon, etc.) and the

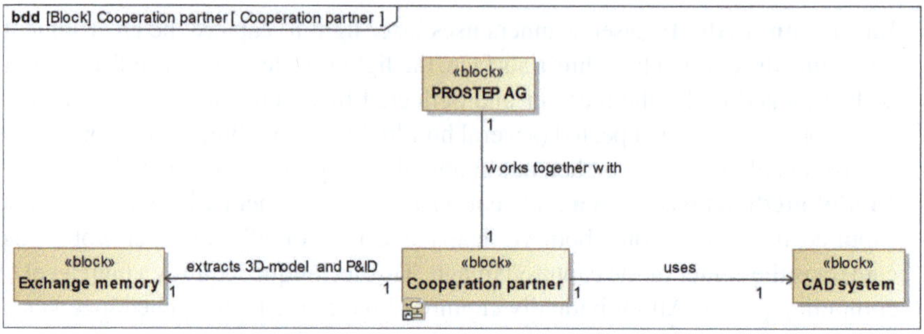

Fig. 6.6 Block diagram cooperation partner

specific working environment. This role is dedicated to a specific partner which is economically independent of each other and acts completely independently in the market. Roles are allocated so that PROSTEP AG performs the vendor-neutral work packages and the cooperation partner (Fig. 6.6) the work packages in a proprietary environment.

All partners considered in Fig. 6.6 govern their commercial terms by a partnership agreement which also defines further common activities (customer acquisition and marketing activities). In addition, the cooperation partner can obtain the CAD model and the P&ID via the exchange memory. The term "exchange memory" describes a user-friendly common data-sharing platform that provides a high-performance exchange of bulk data under the rigorous rules of IT security which is a prerequisite for handling sensitive customer data [31].

The cooperation partner can play various roles. Since the object recognition process relies directly on the scan quality also the number of recognized objects correlates with it. From a scan with less quality, e.g. high amount of fragmentation, also the number of objects recognized correctly will be reduced. In such cases, the know-how and effort of specialists in terms of engineers for process engineering are needed. If the cooperation partner is an engineering office, it has to complete the models by remodeling the not recognized areas.

In other cases, the cooperation partner is specialized in specific CAD vendors and systems respectively. Even though the fundamental understanding of plant engineering is the same each CAD system comes along with specific functionalities or native formats. PROSTEP has development partnerships with the leading vendors on the market. They can take over the role of cooperation partners as well. Native integration into the systems of the vendors is achieved this way. Other PLM vendors can step in here as well. The storage of the digital twin in PLM systems and the enrichment of with additional information can be enabled.

6.4.2 Internal Processes

The internal processes are conducted as an optional extension of the process as described in the source [13] under the opendesc.com service portal [32]. Data conversion experiences over decades brought the capabilities to ensure a generation of digital twins based on the point cloud of a plant. This process can be roughly understood as a conversion of a point cloud to a digital twin.

In order to create a digital twin of a process plant, sales, process planning, software and service development, product management, and managed data exchange are essential capabilities here (Fig. 6.7). The sales department works closely with the cooperation partner in the tasks of marketing and sales. Process planning has to ensure that all customer requirements are implemented flawlessly and without errors using the effects of scale. Based on agile processes, the service and software development are responsible for generating flawless methods and software for creating a digital twin and continuously increasing the reliability of the recognition process [33]. The product manager keeps track of the entire process and assures that each group in the process is doing their work properly and on time. The managed data exchange is in charge of transmitting the data and information from the digital twin in a secure way [34].

If collaboration with the cooperation partner is necessary and desired, both parties mutually require and provide different data and information in order to jointly build the digital twin of the process plant [14]. Since this data is the foundation of the entire process, their main characteristics are summarized in the following Table 6.5.

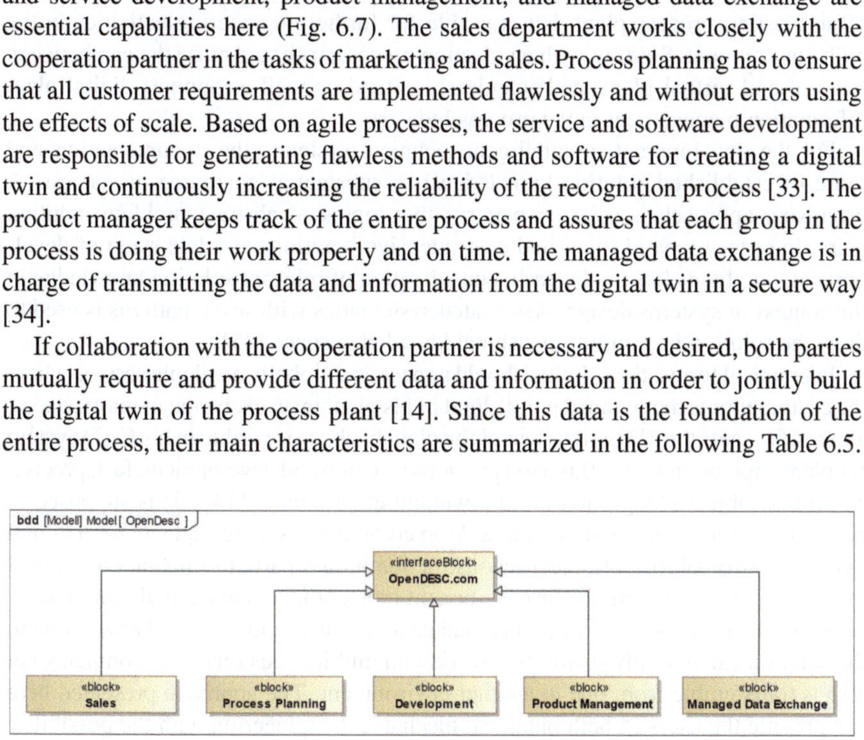

Fig. 6.7 Internal product roles

Table 6.5 Input data for the internal process of PROSTEP AG

Delivery item	Form	Input from	Execution
Point cloud	Scanner-specific project file Neutral format	Capture partner/Plant operator	PROSTEP
P&ID	Neutral or proprietary	Plant operator	Cooperation partner
Plant geometry	Neutral	PROSTEP	PROSTEP
CAD model	Model data	PROSTEP	PROSTEP

This table shows the required data, in which data format and in which form the data must be available as input, as well as the executing instance in the overall process [14].

6.5 Conclusions and Outlook

This chapter intends to imagine a practicable solution approach for the digital twin of an existing process plant that is usable for further engineering activities in the built environment. Starting with an ideation process, it should close the gap between the real and virtual plant, and seamlessly extend as well as support digital plant information management in the process industry.

For the development of intelligent technical systems, the documentation and reuse of established solution knowledge is a substantial guarantee of success. A promising approach is solution patterns that represent abstract problem–solution pairs. In order to avoid time- and cost-intensive iterations at a late stage of development, one focus lies on the early consideration of relevant solution knowledge in the context of systems design. Associated systematics with seven patterns is used to draw the solution in a multidimensional knowledge space [35].

In practical terms, the solution should appear as a globally available service where the customer puts the scanned point cloud of his plant in an exchange memory, and—as a result—gets the 3D digital twin of the plant back, processable in his CAD system for plant engineering [36]. Based on previous research and development, fast, precise, and robust object recognition in a brownfield environment [14, 37] is supposed to be available for implementation on desktop computers as a prerequisite for a digital twin. The affordability of object reconstruction with a partial semantic enrichment must be justified primarily in the cost–benefit ratio compared to manual remastering.

While the processing of the point cloud data runs in a vendor-neutral environment, the solution can not only seamlessly work with multiple customer environments but also is transferable from one to another environment. The approach presented here can provide the users of both plant and mechanical engineering with the possibility to switch to another domain at any time [21]. It fulfills not only the regulatory rules but also supports digital plant information management [38]. Subsequently, dynamic process simulation models can be generated in a standards-compliant way [39].

However, one important drawback must be kept in mind: the dependence on the completeness and the quality of input scan data which is provided by a subcontractor. Although modern scanners provide exact and reliable data, only a few scan errors which can't be identified automatically could compromise the entire project. This indicates that the complicated process of the generation of the digital twin needs to be properly monitored. In further work, an appropriate metric should be defined for the confidence of digitalization of 3D and 2D brownfield plant information [27].

Furthermore, considering the involvement of a multitude of stakeholders, the issue handling should get particular attention. While the current engineering and socio-technical systems were designed to operate in "mostly stable" situations, the sporadic instability and disturbances are at best captured by exception-handling mechanisms, focusing on reliability and robustness [40]. The user needs advice on how to identify undesired events as well as how to correct errors in his model considering collaboration constraints.

References

1. Peruzzini M, Wiesner S (2019) Emergence of product-service systems. In: Stjepandić J, Wognum N, Verhagen WJC (eds) Systems engineering in research and industrial practice. Springer, Cham, pp 209–232. https://doi.org/10.1007/978-3-030-33312-6_8
2. Gustafsson A, Snyder H, Witell L (2020) Service innovation: a new conceptualization and path forward. J Serv Res 23(2):111–115. https://doi.org/10.1177/1094670520908929
3. Sen V (2023) Doing digital the guide to digital for leaders. Business Expert Press, New York
4. Skarka W, Jezierska-Krupa K, Skoberla R (2019) New challenges for ideation in the context of systems engineering. In: Stjepandić J, Wognum N, Verhagen WJC (eds) Systems engineering in research and industrial practice. Springer, Cham, pp 53–88. https://doi.org/10.1007/978-3030-33312-6_3
5. Beverungen D, Müller O, Matzner M, Mendling J, vom Brocke J (2017) Conceptualizing smart service systems. Electron Mark 83(10):1–12. https://doi.org/10.1007/s12525-017-0270-5
6. Kohli R, Melville NP (2019) Digital innovation: a review and synthesis. Inf Syst J 29(1):200–223. https://doi.org/10.1111/isj.12193
7. Gimpel G (2020) Bringing dark data into the light: illuminating existing IoT data lost within your organization. Bus Horiz 63(4):519–530. https://doi.org/10.1016/j.bushor.2020.03.009
8. Baltuttis D, Häckel B, Jonas CM, Oberländer AM, Röglinger M, Seyfried J (2022) Conceptualizing and assessing the value of internet of things solutions. J Bus Res 140:245–263. https://doi.org/10.1016/j.jbusres.2021.10.063
9. Wang K (2019) Towards a taxonomy of idea generation techniques. Foundations of Management 11:65–80. https://doi.org/10.2478/fman-2019-0006
10. Kirjavainen S, Hölttä-Otto K (2020) Deconstruction of idea generation methods into a framework of creativity mechanisms. In: Proceedings of the ASME 2020 international design engineering technical conferences and computers and information in engineering conference. Volume 8: 32nd international conference on design theory and methodology (DTM). Virtual, Online. 17–19 Aug 2020. V008T08A004. ASME. https://doi.org/10.1115/DETC2020-22038
11. Herring SR, Jones BR, Bailey BP (2009) Idea generation techniques among creative professionals. In: Proceedings of the 42nd Hawaii international conference on system sciences. https://doi.org/10.1109/HICSS.2009.241
12. Peruzzini M, Wiesner S (2019) Systematic development of product-service systems. In: Stjepandić J, Wognum N, Verhagen WJC (eds) Systems engineering in research and industrial practice. Springer, Cham, pp 265–296. https://doi.org/10.1007/978-3-030-33312-6_10
13. Stjepandic J, Sommer M, Denkena B (2022) DigiTwin: an approach for production process optimization in a built environment. Springer International Publishing Switzerland. https://doi.org/10.1007/978-3-030-77539-1
14. Kremer P, Lützenberger J, Müller F, Stjepandić J (2022) An approach for the incremental update of a digital twin of a process plant. Advances in Transdisciplinary Engineering 28:310–319. https://doi.org/10.3233/ATDE220660
15. Bondar S, Hsu JC, Pfouga A, Stjepandić J (2017) Agile digital transformation of system-of-systems architecture models using Zachman framework. J Ind Inf Integr 7:33–43. https://doi.org/10.1016/j.jii.2017.03.001
16. Bondar S, Ruppert C, Stjepandić J (2014) Ensuring data quality beyond change management in virtual enterprise. International Journal of Agile Systems and Management 7(3/4):304–323. https://doi.org/10.1504/IJASM.2014.065346

17. Moon D, Chung S, Kwon S, Seo J, Shin J (2019) Comparison and utilization of point cloud generated from photogrammetry and laser scanning: 3D world model for smart heavy equipment planning. Autom Constr 98:322–331. https://doi.org/10.1016/j.autcon.2018.07.020
18. Agapaki E, Brilakis I (2020) CLOI-NET: class segmentation of industrial facilities' point cloud datasets. Adv Eng Inform 45:101121. https://doi.org/10.1016/j.aei.2020.101121
19. Kuhn O, Liese H, Stjepandić (2011) Methodology for knowledge-based engineering template update. In: Cavallucci D, Guio R, Cascini G (eds) Building innovation pipelines through computer-aided innovation. Springer, Berlin, pp 178–191. https://doi.org/10.1007/978-3-642-22182-8_14
20. Sommer M, Seiffert K (2021) Scan methods and tools for reconstruction of built environments as basis for digital twins. In: Stjepandić J et al (eds) DigiTwin: an approach for production process optimization in a built environment. Springer Nature, pp 51–78. https://doi.org/10.1007/978-3-030-77539-1_4
21. Grau M, Korol W, Lützenberger J, Stjepandić J (2021) Automated generation of a digital twin of a process plant by using 3D scan and artificial intelligence. Advances in Transdisciplinary Engineering 16:93–102. https://doi.org/10.3233/ATDE210087
22. Stjepandić J, Sommer M (2022) Object recognition methods in a built environment. In: Stjepandić J, Sommer M, Denkena B (eds) DigiTwin: an approach for production process optimization in a built environment. Springer series in advanced manufacturing. Springer, Cham, pp 103–134.https://doi.org/10.1007/978-3-030-77539-1_6
23. Sommer M, Stjepandić J, Stobrawa S, von Soden M (2023) Automated generation of digital twin for a built environment using scan and object detection as input for production planning. J Ind Inf Integr 33:100462. https://doi.org/10.1016/j.jii.2023.100462
24. Cheng L, Wei Z, Sun M, Xin S, Sharf A, Li Y, Chen B, Tu C (2020) DeepPipes: learning 3D pipelines reconstruction from point clouds. Graph Models 111:101079. https://doi.org/10.1016/j.gmod.2020.101079
25. Lee ID, Lee I, Han S (2021) 3D reconstruction of as-built model of plant piping system from point clouds and port information. Journal of Computational Design and Engineering 8(1):195–209. https://doi.org/10.1093/jcde/qwaa072
26. Oh I, Ko KH (2021) Automated recognition of 3D pipelines from point clouds. Vis Comput 37:1385–1400. https://doi.org/10.1007/s00371-020-01872-y
27. Sierla S, Azangoo M, Rainio K, Papakonstantinou N, Fay A, Honkamaa P, Vyatkin V (2022) Roadmap to semi-automatic generation of digital twins for brownfield process plants. J Ind Inf Integr 27:100282. https://doi.org/10.1016/j.jii.2021.100282
28. Sattler G, Kasper W (2000) Verfahrenstechnische Anlagen: Planung, Bau und Betrieb. WILEY-VCH Verlag, Weinheim
29. Wossog G (2008) Rohrleitungsbau, 3rd edn. Vulkan-Verlag, Essen
30. Holt J (2010) Modelling enterprise architectures. The Institution of Engineering and Technology, Stevenage
31. Stjepandić J, Sommer M, Stobrawa S (2022) The commercialization of digital twin by an extension of a business ecosystem. In: Stjepandić J et al (eds) DigiTwin: an approach for production process optimization in a built environment. Springer Nature, pp 205–233. https://doi.org/10.1007/978-3-030-77539-1_10
32. NN (2023) https://3digitaltwin.opendesc.com/. Accessed 15 Jan 2023
33. Karagiannis D, Robert Andrei Buchmann RA, Utz W (2022) The OMiLAB digital innovation environment: agile conceptual models to bridge business value with digital and physical twins for product-service systems development. Comput Ind 138:103631. https://doi.org/10.1016/j.compind.2022.103631
34. Biahmou A, Stjepandic J (2016) Towards agile enterprise rights management in engineering collaboration. International Journal of Agile Systems and Management 9:302–325. https://doi.org/10.1504/IJASM.2016.081564
35. Anacker H, Dumitrescu R, Kharatyan A, Lipsmeier A (2020) Pattern based systems engineering—application of solution patterns in the design of intelligent technical systems. Proceedings of the Design Society: DESIGN Conference 1:1195–1204. https://doi.org/10.1017/dsd.2020.107

36. Sierla S, Azangoo M, Fay A, Vyatkin V, Papakonstantinou N (2020) Integrating 2D and 3D digital plant information towards automatic generation of digital twins. In: 2020 IEEE 29th international symposium on industrial electronics (ISIE). https://doi.org/10.1109/ISIE45063.2020.9152371
37. Sierla S, Pourakbari-Kasmaei M, Vyatkin V (2022) A taxonomy of machine learning applications for virtual power plants and home/building energy management systems. Autom Constr 136:104174. https://doi.org/10.1016/j.autcon.2022.104174
38. Lacroix S, Ostermeyer E, Le Duigou J, Bornard F, Rival S, Mary MF, Eynard B (2023) Lessons learnt in industrial data platform integration. Procedia Computer Science 217:1660–1669. https://doi.org/10.1016/j.procs.2022.12.366
39. Xavier M, Håkansson J, Patil S, Vyatkin V (2021) Plant model generator from digital twin for purpose of formal verification. In: 26th IEEE international conference on emerging technologies and factory automation (ETFA), pp 1–4. https://doi.org/10.1109/ETFA45728.2021.9613704
40. Ramezani J, Camarinha-Matos LM (2020) Approaches for resilience and antifragility in collaborative business ecosystems. Technol Forecast Soc Chang 151:119846. https://doi.org/10.1016/j.techfore.2019.119846

[illegible]

Chapter 7
Implementation of a Digital Twin of a Process Plant

Abstract In this chapter, we describe how to adopt different software components based on the previously presented conceptual solution and generate an integrating digital twin for a business workflow as a service extension. The focus of this software system lies in the implementation of methods from computer vision that can reliably recognize the existing objects in a process plant with their structure and interconnections. This intention benefits from numerous methods which were developed in the past years to tackle the challenge of the recognition of 3D objects under various conditions. A brief review of such methods is presented here, in particular concerning difficult environmental impact (vapor, dust, smoke, darkness, and dirt). Methods are distinguished according to data input: image, point cloud, or video. For several reasons, the implementation of an automatic object recognition procedure is realized by using existing convolutional neural networks, also known as deep learning. Literature review shows that effective and versatile automation capabilities of deep learning combined with large-scale processing may be an adequate means for the challenges of the extent and complexity of a process plant. Further is the recognition procedure described in more detail, in particular how the piping system is built up in its full complexity coming from singular components. This recognition runs iteratively in four steps with a mutual interdependence. The entire point cloud is processed by segmentation, where the pipe system is extracted. In two subsequent steps (clustering and classification) the point cloud is subdivided further to recognize the singular parts. While piping systems consist of forked structures with complex configurations and partial occlusion, the exact recognition of piping centerlines is conducted based on the position of singular parts. A robust clustering and graph-based aggregation yield a coherent pipe model before it is linked with piping and instrumentation diagram to the digital twin. Due to the high complexity and variance of the tasks, some individual tasks must run fully manually or with partial human assistance. Finally, we present how the singular steps are automated and orchestrated by using a workflow automation platform. Our workflow promises good results on pipe models with varying complexity and density both in synthetic and real cases.

J. Stjepandić et al., *Generation and Update of a Digital Twin in a Process Plant*,
https://doi.org/10.1007/978-3-031-47316-6_7

Keywords Digital Twin · Object recognition · Convolutional neural network · Deep learning · Point cloud · Workflow automation · Piping and instrumentation diagram

7.1 Introduction

Implementation represents here the realization of the previously defined conceptual solution and processes in a software system, taking into account framework conditions, rules of the software development process, and targets, as defined in the specification of the system. The aim of this software system is to generate the digital twin of a process plant using a point cloud of it which was previously acquired by a scanner or an alternative device. This point cloud should be transferred to a commercial system for digital asset management. Existing objects in a plant and their interconnections are provided by a key module of this software system providing object recognition. Due to the high complexity of a plant, the diversity, and the multitude of built objects, the generation procedure is considered highly automated with low manual assistance with sufficient operation protocols which facilitate error tracking. The software system builds on the methods, structures, and software modules that were already proven in previous work [1].

Computer vision as an interdisciplinary scientific field involves using computers to gain a detailed understanding of visual data, which is a similar approach to that of human visual systems. It is used to enhance the tasks of industrial engineering and management by enabling the acquisition, processing, analysis of digital images, and the extraction of high-dimensional data from the real world to produce information to improve decision-making. Furthermore, computer vision can provide practitioners with rich digital images and videos (e.g., location and behavior of objects/entities, and site conditions) about a project's prevailing environment and therefore enable them to better manage the construction progress [2]. Computer vision is used to examine specific issues in the industry such as tracking people's movement, progress monitoring, quality and productivity analysis, process, health and safety monitoring, and postural ergonomic assessment [3]. A plethora of similar applications are also used in medicine and defence [4, 5].

Coming from the wider field of computer vision, object recognition has gathered even more research contributions in the past years due to its growing applications including robot vision, autonomous driving, security, quality assurance, consumer electronics, human–computer interaction, content-based image retrieval, intelligent video surveillance, and augmented reality [6]. Object recognition aims to determine whether an instance of an object from a specific category (human, animal, car, or machine) is available in a scene, to identify objects in a scene, and to estimate their position with sufficient accuracy. An important paradigm of object recognition is to first define suitable surface representations, offline, for 3D objects and save those representations in a database [7]. During online recognition, a similar representation of a scene is matched with the representations stored in the database to recognize objects which are present in the scene [8]. The main challenges associated with surface representation and 3D object recognition are occlusions caused by the presence of multiple objects in the scene, clutter due to unwanted and unordered objects (e.g., job sites), and robustness to noise and resolution [9].

To contactless acquire the geometrical shape of an object, a multitude of technologies and devices is available. On active technologies, the surface of an object or a complete environment is captured by illuminating them using an additional source of radiation (e.g., a laser beam). A 3D profile is generated, as seen from the perspective of the point of measurement. In passive technologies, a sensor picks up the reflected or emitted radiation from the surface of an object in its normal environment. In an additional step, the 3D structure is inferred by interpreting the depicted scene [10].

Active technologies (e.g., laser scanners) can achieve high accuracy and a large range. Due to its complexity of operation, scanning with a laser scanner still requires expert knowledge. On the passive technology (e.g., camera), on-site recording time is less than in the active case by a factor of three. The resolution of the scan can be easily adjusted, the more images from different angles of the scene are taken, the higher the resulting resolution of the scan. A large disadvantage is, that the object to be scanned needs to have distinctive and prominent features visible throughout individual images to correctly assess the connections between them. Changing lighting conditions during the measurement have a negative effect on the reconstruction [10]. This and further possible negative environmental impacts (dirt, vapor, dust, smoke, darkness) have caused the laser scanner to be seen as the most reliable, efficient source of input data for object recognition with high environmental adaptability.

More recently, smartphones were upgraded with LiDAR sensors and corresponding apps. For industrial applications, robust laser scanner devices which mostly accompany an auxiliary high-resolution camera, are available for years and can be seen as the standard source of input data for the recognition process. A further advantage of the laser scanner poses in its fast, target-free, and fully automatic self-calibration process using the isomorphism constraint and ambiguity judgment algorithm [11]. This plethora of acquisition devices is rounded off by a vast amount of mobile augmented reality (AR) devices whose development is driven by gaming technology. Further applications of scanned 3D data comprise progress tracking, dimensional compliance control, and the location, movement, and assembly of materials [12].

Leveraging the availability of high-performance laser scanning systems, data processing of large point clouds has in parallel fashion progressed with a significant speed. Intense activity in the development of automated data processing algorithms, especially in the field of object recognition [13], has been noticed. The emergence of these algorithms is related to an intense demand for engineers and technicians from different industries and domains. Different applications have in common that they all demand an efficient conversion of the raw laser scan point cloud data to native geometrical 3D models, but each with context-specific requirements. Concerning the evolution of laser scanning technology, a broad overview of state-of-the-art algorithms, different best practices, and most recent processing tools such as the emergent application of deep learning to dense point clouds is elaborated by Rivero and Lindenbergh [14].

Finally, the key task to be implemented consists of how to transform semantic unstructured or low-structured digital data (e.g. a high-quality image, video, or point cloud) of a process plant in a compact, process-oriented digital representation that can be seamlessly used for generation of Digital Twin [15, 16]. Like a 3D search, this procedure uses classification and segmentation techniques to assign both class-aware labels with respect to defined object classes and instance-aware labels with respect to objects in the considered scene. In particular, object recognition comprises 3D object reconstruction, 3D object localization, and 3D modeling based on the outcome of data acquisition [17].

For humans, the visualization of a dense point cloud already allows reasoning about the occurrence of specific objects in a scene. A diversity of objects such as buildings, ground, tanks, pipelines, or valves can easily be detected by solely considering the spatial arrangement of 3D points. Furthermore, the human capability to detect a diversity of objects is rather robust to occlusions, strongly varying point density, and irregular point sampling. Accordingly, it seems desirable to adopt such capabilities through automated systems that uniquely assign each 3D point of the point cloud with a corresponding structural label. Additionally, a respective approach should also be capable of coping with local variations in point density and irregularly distributed 3D points [18].

The generic process of object recognition is illustrated in Fig. 7.1. At first, an object with distinctive physical properties must be acquired by an aforementioned observation device. Recognition methodology will be described in the following section. In the next step, the features which can be identified in the scene must be classified against an object patterns database. All recognized features are integrated with the next step to objects by using external features like 3D search. By using environment knowledge for distinctive characteristics, the result of this process is

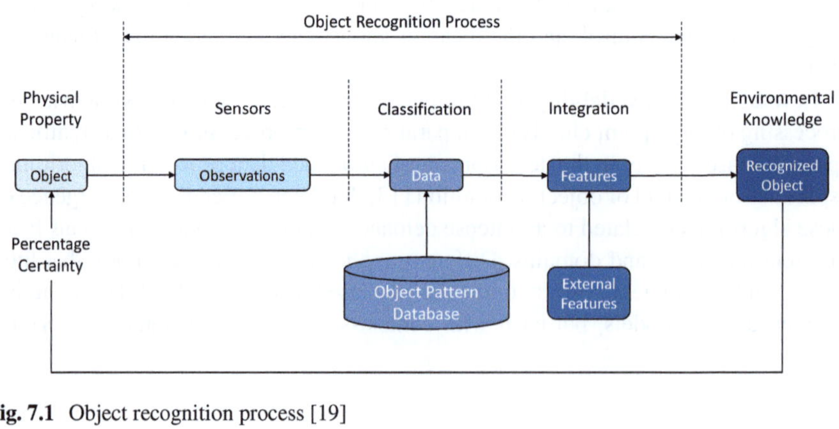

Fig. 7.1 Object recognition process [19]

the recognized object, based on a recognition accuracy to distinguish between two similar objects [19].

If the steps of classification and integration in Fig. 7.1 are merged, then the procedure is applied by searching for and recognizing the entire objects rather than the individual features which would be collected to an object further. In the practical implementation, this step is realized by identification of a geometric object from a library which is derived from CAD models of plant components, instead of identifying the detail feature of each object. Simultaneously, localization is conducted by an appropriate auxiliary geometrical approach (e.g., calculation of a bounding box). Conversely, while the 3D search works like a similarity check, a library of all appearing objects of interest with all distinguishing variants is a prerequisite that this approach will be successfully completed [20].

While a plant model comprises much more information than the full list of its components, this implicit information which basically consists of the pipeline structure must be appropriately added to build the digital twin. For this purpose, secondary information must be either determined or derived. The most important is the connectivity in a pipe string which aims at the reconstruction of flows and processes in a piping system. At the end of the object recognition workflow, a seamless, full list of components in the related piping system is necessary which can be imported into a commercial CAD system (e.g., AVEVA E3D) without further rework.

As in the previous work practiced [1, 20], algorithms based on convolutional neural networks (CNN)—known also as "deep learning"—, a subcategory of machine learning, were selected to implement the object recognition in all stages of generation of the digital twin. The main reason is the popularity of machine learning and CNN in particular in the area of computer vision which yields the development of a multitude of new object recognition frameworks based on standard machine learning platforms and contemporary procedures like MLOps (Machine Learning Operations) [21] which are mostly available in open-source repositories.

The CNN frameworks can process the disordered and discrete point cloud data directly and construct some regular-like operators in the deep learning method [22]. These methods of processing point cloud data have been considerably facilitated in a variety of ways. The supervised learning-based approaches generally have wide applicability because a high accuracy can be achieved at a viable expense. Furthermore, a considerable amount of standard data for training are already available [23, 24] along with recommendations on how to improve the predicted labels by enforcing post-processing rules.

The remainder of this chapter is structured as follows: in Sect. 7.2, the methodology for object recognition of the structural model is presented by distinguishing between image, point cloud, and video-oriented methods. In Sect. 7.3, the object recognition procedure is explored based on supervised learning. The implemented procedure for object recognition is described in Sect. 7.4, followed by the description of the practical workflow in Sect. 7.5. Together with future perspectives, it is discussed in Sect. 7.6. Finally, Sect. 7.7 summarizes the conclusions and outlook.

7.2 Methodology for Object Recognition

Methodology for object recognition can be classified on various criteria [20]. The most impactful criterion is the input data type. In our case, it is a point cloud of the process plant in question [25]. Alternative inputs are images [26] or video [27] of a scene. As will be presented further, object detection as such heavily relies on the input data type. However, methods for extraction of point clouds from videos or derivation of point clouds from images can be used to extract point clouds as a universal data type [20]. This is, in particular, important because the "point cloud of a plant" is rather a composition of multiple point clouds, matched with each other during a procedure called registration, than a monolithic object. The inputs can come not only from different devices but also from different data types. This can yield fluctuating consistence and quality of the point cloud and, subsequently, lead to varying results. However, such an issue can be resolved by repeating the scan for the portion of the plant in question.

7.2.1 Image-Oriented Methods

With the image-oriented method, the objects are extracted from the individual pixels of the image. In the example of a color camera, each pixel represents a recording of the light. The pixel contains information about color and brightness and these in turn depend on the angle and lighting, as well as the reflection of an object. For object recognition, data acquisition of the objects to be recognized is required. An image can be recorded in various ways (e.g. color, X-ray, or infrared cameras). With object recognition, the objects are recognized with bounding boxes and then assigned to specific types or classes.

Object recognition is divided into two areas. On the one hand in the classification and on the other hand in the object localization. In the classification, the object is assigned to a class or a type. With object localization, on the other hand, the detected object is searched for in an image, and its position is determined by a bounding box [28].

Image-oriented methods use model-based methods and methods based on appearance. With model-based recognition, the object is recognized based on the contours. An edge is a strong change with regard to color or brightness values. This way, strong changes lead to an intense edge. Further, the changes are analyzed, and the object is thereby identified and recognized. With the methods based on appearance, on the other hand, the distribution of intensity values of an object is calculated for histograms. The course of the histogram is described by the respective object like a template. These histograms are calculated for object recognition and compared with the templates available in the object database. An object is recognized when the histogram matches the template [29].

7.2.2 *Video-Oriented Methods*

In the video-oriented method, a video is defined as a sequence of related images. This describes the object to be recognized, whereby the images of the description cannot be analyzed individually. In the case of a video, the temporal component is added. The analysis of object recognition takes place in two areas: firstly, from the well-known area of image-based object recognition, and secondly by analyzing the change in an object over the individual images. The objects do not only have been recorded by one camera but by several cameras from different positions and angles. After that, the image sequences from the individual cameras must be linked and synchronized in order to obtain a uniform image. In particular, video salient object detection emphasizes objects that draw attention in a video, and this has emerged as an active part of research due to its increasing attention on applications like object tracking, object segmentation, and action recognition. Utmost all existing approaches calculate the salient object motion using optical flow since it is highly sensitive to variation in illumination and changes in localization [30].

7.2.3 *Point Cloud-Oriented Methods*

Another approach for detecting 3D objects is the point cloud-oriented method. Here, the (2D) image-oriented method is expanded with the component of depth information. There are various approaches for doing this, which are briefly explained below [8]:

- *Feature-based method*: This method is the most commonly used type of evaluation. The 3D objects are described with local and global features. To enhance object recognition, both approaches (global and local) are combined. With the classification into a category by global features, local features for the recognized category can explicitly be used for this area. The result of the analysis of the local features describes the 3D object completely.
- *View-based method*: This method captures the entire object in its appearance. E.g. it can be the outer edge of a machine, which can be different from every perspective. Object recognition means that an object cannot be recognized from an unfavorable viewing angle.
- *Graph method*: With this method, several properties and their connections to one another are described. The graph describes the object to be recognized by the individual dependencies between the properties, while they are represented by nodes, and the connections are defined as edges (directed and non-directed). The different properties are classified and arranged in a graph.

7.3 Object Recognition Procedure

CAD and mechanical models are predominantly composed of repetitive basic structures to facilitate easy and economic fabrication. Surface reconstruction involving the local fitting of primitive structures has long been the standard in reverse engineering. Furthermore, piping system reconstruction needs to capture the skeleton and topology. In contrast to other thin structures, pipelines have a specific cylindrical nature while lacking regular patterns such as fences. Furthermore, they are typically rigid bodies in contrast to e.g. flexible wires and their industrial significance demands a highly accurate result. Industrial plants are generally assumed as assemblies of mechanical parts. In focus lie parts such as pipes, elbows, flanges, tees, and crosses. The raw scan of a pipeline is input, and its part-based reconstruction poses the output [23].

The kernel of the object recognition procedure is based on the procedure for the identification of pipe systems (Fig. 6.2). To determine the piping system as such, all its components must be detected, and the semantics of the system rebuilt based on the detected interconnections and design rules for pipe systems (Sect. 4.3). In a first step, the entire point cloud provided by a scanner is prepared for the semantic segmentation. In the second step, the point cloud which represents the process plant or a part of it is segmented into several dominant object categories (pipeline, equipment, structure, etc.). Singular categories are segmented further into clusters that comprise singular objects. At the end of this process, each recognized object is assigned to a known object from the CAD library.

Since the entire object recognition process is based on CNN, the implementation here means rather the chaining of already available, individual CNN frameworks and optimization of training procedures than the development of entirely new algorithms and software. Conversely, usage of CNN requires experience in model preparation, validation, and interpretation of results in order to obtain the optimal process parameters (e.g., allowed tolerances).

7.3.1 *Learning Procedure*

Following the main trends in research [31–33], CNN based on supervised learning is selected for object recognition because it can cover both regression and classification. It provides quite accurate and reliable results since the model is trained and the labels signify the data they belong to. Figure 7.2 shows the significant supervised learning model steps. Supervised learning data is split into three categories: training set, validation set, and test set. The training set is used to train the model, and the validation set is used to tune in the model parameters and improves the model's credibility. This process is continued until the model provides good results [34]. For the recognition of piping components, a dozen of plants with different modules to achieve a sufficient level of learning are required.

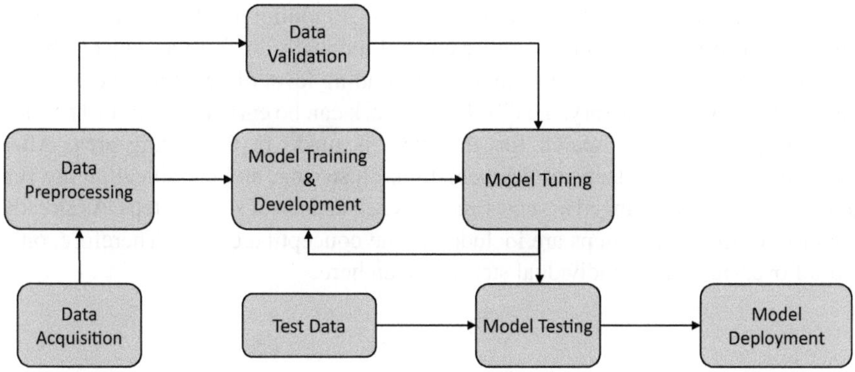

Fig. 7.2 Procedure of supervised machine learning [34]

A manual effort to prepare dedicated portions of scanned point clouds for training is significant and has a crucial impact on the quality of the final results. After this, the model is tested on test data, which the model has never analyzed before. These sets of results are perceived to be results that the model will give once deployed in the real world. These models are trained with the existing datasets and used to determine the new data results. It is an excellent example to project machine learning capabilities in projecting the results after learning from the previously analyzed data [34]. In addition, the available frameworks comprise a fully connected layer followed by dropout to eliminate possibly the same interpretation of prediction that can help to avoid overfitting [35].

For users, overfitting is rarely obvious, but its impact is ubiquitous. For recently published CNN frameworks a comparable performance is announced according to standard data sets. If predicted results and indicators are sensitive to change (e.g., input data quality), this would imply strong overfitting. The attempts to reduce overfitting lie in the extension of a standard test base by more model variants for each object category. For fittings and equipment, just as important is the compromise between the level of detail, which takes local features into account, and limited model complexity for each object in the object library. This can be conducted by compilation of existing libraries (e.g. provided by customers) [36].

Since only supervised learning is used here, the methods for each step during recognition basically consist of the tasks of "training" and "testing". Data validation is conducted in the early phase only. The complete process of object recognition is divided into the steps of segmentation, clustering, and classification, which follow in a serial order whereby each step can be further subdivided. So, the segmentation of pipe systems could be subdivided into rigid and flexible pipes or the main and auxiliary pipe systems. The disadvantage of this method is that any early error propagates to the overall result.

Training: With regards to the use of deep learning specifically the so-called "supervised learning (training)" is meant. It consists of a predetermined task to be learned whose results are known. Consequently, the results of the learning process can be compared with the known, correct results, i.e., "monitored". In the case of three-dimensional objects, a data set, either created manually or from an object library, must be available for each object in a corresponding level of detail in order to carry out the training. If necessary, the CNN framework can be enriched with further data.

Testing: The actual search for an object is made in the testing area. After the training, the three steps of segmentation, clustering, and classification are run through, which are arranged in stages and can each consist of several steps. As already mentioned, the testing steps are included in the conceptual design. Therefore, only a brief overview of the individual steps is given here.

7.3.2 Deep Learning Pipes

During the design of piping systems, pipe modules are composed of pipe components and pipe support elements. In the context of the reconstruction of a process plant, the focus lies on pipe components because pipe support needs to be replaced or moved after the final course of the pipeline is defined. Further, parts and supports are not considered such as floors, fences, etc. due to the problem's magnitude and the low impact on digital twin [23].

A primitive-based segmentation method for mechanical CAD models can detect the most prismatic parts. The method assumes a limited number of dominant orientations that primitives are either parallel or orthogonal to, narrowing down their search space. Finally, they generate an over-complete set of primitives and formulate the segmentation as a set cover optimization problem [23].

The recognition of components and reconstruction of the piping system run gradually in a bottom-top manner processing the entire space. Singular components as depicted in Figs. 2.2, 2.3, and 2.4 are sought for each category in the available space. Possible connections are also identified in multiple variants. Outcomes are a part list and a structured data model which describes the content of the space in an abstract way.

In the last step, formalized objects are replaced by the actual 3D part models, and the scene is reconstructed. For fine-tuning, the parts fitting is readjusted by an algorithm employed to minimize the difference between two clouds of points and transform them to better fit the point data [23].

7.4 Implemented Procedure

Apart from the overall objectives in software engineering, the implementation of a process chain for object recognition addresses three mutually competing objectives: quality, efficiency, and robustness. High-quality recognition has to accurately localize and recognize objects in point clouds in their right dimensions without wrong recognitions (e.g., a valve instead of a pump) considering their almost endless variety. High efficiency requires the entire detection task to run at a sufficiently high frame rate with acceptable memory and storage usage. In addition, it should also be considered that the process starts with a huge amount of data, measured in hundreds of gigabytes, and the digital twin is delivered in a common and workable model volume. In a certain contrast to quality and efficiency, robustness refers to the ability to handle not complete and erroneous data [20].

7.4.1 Segmentation

With segmentation, the entire point cloud is divided into individual segments. When focusing on such a semantic classification task, the number of defined object classes and the similarity between the classes typically play important roles, and the involved features should be sufficiently distinctive to allow for distinguishing between the defined object classes [18]. These segments correspond to specific functionality (e.g., a pipe system). In a point cloud, each segment is labeled with a specific color (Fig. 7.3).

For each point, it is determined whether it belongs to an object or not. This assignment creates a binary image. If a point belongs to an object, the value of that point becomes one, otherwise, the value becomes zero. By dividing the point cloud into regions, the objects can be recognized after segmentation. There are three basic segmentation options:

Fig. 7.3 Example segmentation [37]

- Point- or pixel-oriented segmentation, which only uses the gray values of the individual points/pixels.
- Region-oriented methods use gray values in regions.
- Edge-based methods run along the edges of an object once they have been detected.

During the segmentation point clouds are better structured. Thereby, those objects that are of particular interest can be preselected from an unmanageable point cloud and the data volume for further processing is drastically reduced by a factor of 1.000. The complexity of a use case increases with the number of existing segments. Segmenting the complete pipe system gets particular importance because, in the subsequent steps, the relations between components need to be recognized. Any errors in the selection are to be considered critical for the whole process.

7.4.2 Clustering

In general, clustering is the partitioning of a dataset into different classes or clusters according to a particular criterion, such that the data objects are as similar as possible and at the same time as different as possible for data objects that are not in the same cluster. Clustering aims to assign the resulting point clouds from the segmentation to specific spatially compact subclasses (clusters) that contain one or more objects that are still to be recognized. That is, after clustering, data of the same class are brought together as much as possible, and different data are separated as much as possible. The basic idea is that the neighboring elements (e.g., pipe segments) are combined into a cluster in a point cloud based on cluster conditions.

In clustering, the points in the cloud are clustered together. In this step, each segment is searched for the cluster of points. If a self-contained region is found, it is enclosed in a bounding box. It encloses the maximum extent of an object. This determines the position with the x, y, and z coordinates as well as the orientation of the object in space.

As an example, Fig. 7.4 shows the clustering of the pipe system in a biogas plant [37]. Different clusters are distinguished by different colors of the corresponding point clouds.

7.4.3 Classification

The final object recognition (recognition in the narrow sense) takes place in the so-called classification. It is based on certain statistical methods that calculate the probability that a certain object is contained in a cluster in a certain appearance form. Object structure with the individually recognized objects is the result and can be the output in the form of a structure file with all objects. This file contains the x, y, and z coordinates as well as the orientation of an object defined by a spatial transformation.

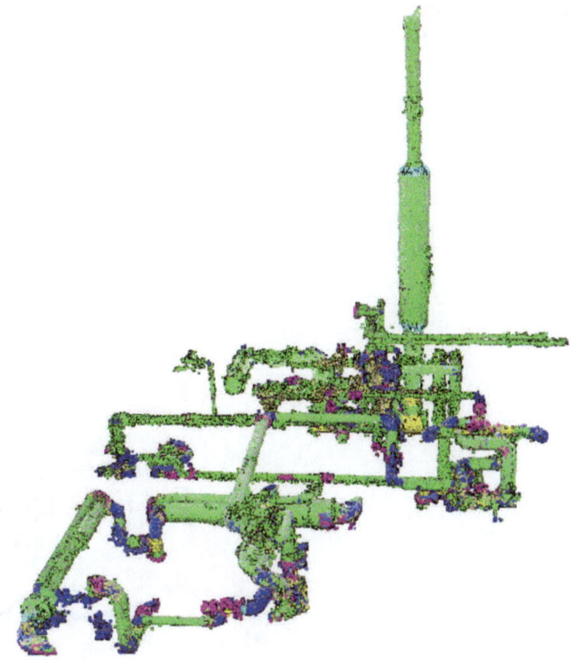

Fig. 7.4 Example clustering [37]

Thus, the object is clearly positioned in space. Depending on the object category, further processing is arranged. For an object in the category of equipment, CAD models can be taken from a library at the desired level of detail and inserted into a CAD system. The objects have the correct position and the correct orientation in space.

For piping systems, which lie the focus of the entire process, an alternative procedure is selected because the design, in the first step, primarily needs the piping system configuration and the diameter. As an illustration, different piping segments are ordered according to their diameter and illustrated in different colors in Fig. 7.5. When all components of a piping system are recognized they can be structured and set in relation to each other. In this view, the user can easily make a preliminary visual check of the results.

Based on the preliminary results of the recognition of components, a robust determination of system configuration can be executed (Fig. 7.6) because this is the fundamental information ("backbone") for the design of piping systems based on the adjacent segments, fittings, and components. This procedure takes several criteria into account: continuous centerline, continuous pipe diameter, etc. While fittings are dimensioned according to the pipe diameter, they can be placed more precisely based on the almost exact course of centerlines [38].

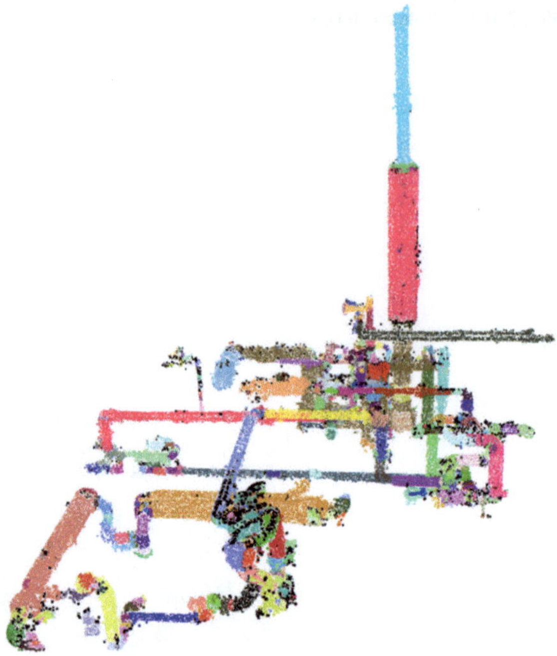

Fig. 7.5 Example classification [38]

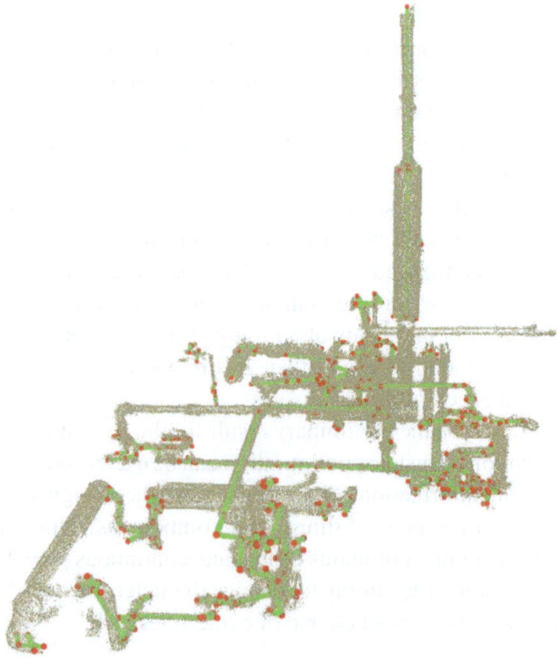

Fig. 7.6 Example detection of centerlines [38]

7.5 Practical Workflow

The process chain described in this chapter consists of a few dozen of working steps that run or are affected by (at least) four legally independent parties. Therefore, a complex, concatenated procedure described in Sect. 7.4 need to be monitored, tracked, and controlled accurately. Before such an implementation by an appropriate workflow software, it must be defined how the process steps scan, segmentation, classification, and linking with P&ID (Fig. 2.6) interact with each other.

Due to the high complexity and variance of the tasks, some individual tasks must run fully manually or with partial human assistance. Supposed a sufficient quality of the point cloud of the object in question (entire plant, unit, or a part of it), the task scan can be omitted from consideration here. Nonetheless, careful visual control of the input point cloud by an experienced user is indispensable because many, spatially and contextually independent errors may occur, mostly due to the course of the scanning process [20].

While this service is offered to customers from the industry who expect a precisely defined individual scope of delivery, especially regarding time and quality, this workflow has to meet high requirements for stability and robustness of the entire process for each order [39]. Therefore, process monitoring, error tracking, and documentation get particular importance.

7.5.1 Interactive Model Preparation

Usually, the output of the scan process is stored in a large, vendor-specific project file which comprises not only the desired point cloud but also control, high-resolution images, the position of the scans, and further information [26]. At the moment, such a point cloud cannot be used immediately for object recognition for several reasons [18]. It may contain false or undesired information, e.g. surroundings of the plant or reflections by the large glass or glittering surfaces. Therefore, point cloud preprocessing in an interactive tool (e.g., point cloud editor) is a prerequisite to extracting a point cloud model for seamless further processing, e.g., segmentation, clustering, and classification [20].

Visual check of the raw cloud is indispensable for the prevention of unnecessary consequential errors, e.g., if the automatic registration has failed. Besides, the data set must be suitably positioned in space. This happens by selecting an appropriate reference coordinating system. Due to the high noise levels in industrial plant scans, some hand-crafted features may appear heuristically. Therefore, the model clean-up, the noise filtering, and the outlier removal should be conducted concomitantly during these interactive steps [20].

Eventually, the entire point cloud must be split, e.g., to extract the portion of interest which will be processed further. This interactive capability gets particular importance in case of an (incremental) update of a process plant or a unit [38] which is a frequent use case. The extraction of models and portions for (additional) training in segmentation respectively classification happens at this milestone. Finally, the point cloud must be exported in the desired format (e.g., e57) for processing in CNN frameworks [20].

From this point on, the process runs automatically except for training activities which will continuously decline in their importance and extent the more data sets are already trained. While an increasing multitude of components in a plant cannot be expected, the additional training will be limited and supported by the availability of further test bases [24, 25].

Mesh reconstruction and point cloud segmentation are encapsulated in separate utilities for the object recognition framework. A user front-end with an integrated viewer was developed to summarize all automatic running steps and monitoring functions. It controls the steps of semantic segmentation, cluster extraction, cluster classification, and determination of centerlines. If necessary, singular steps can be repeated [20].

A point cloud model of an entire plant comprises data volume in the range of several hundreds of gigabytes up to terabytes of memory space which is challenging for both humans and machines as well as extremely time-consuming because some steps must be done interactively. The preparation steps also aim to reduce the volume of data to an extent that is much easier for the interactive work and lowers the demand for hardware [20].

Using the aforementioned methods and utilities, a reduction of the data volume of the point cloud by a factor of more than 1000 becomes possible without losing the quality of the content. The range of volume reduction heavily depends on the desired object recognition framework [20].

7.5.2 Workflow Automation

Increasing demands for system complexity and performance require tight coordination between disciplines as well as more interoperability between the supporting software tools. For a productive environment, a formal description of a process must be transformed into a workflow. Workflow is understood as the sequence of steps involved in advancing from the beginning to the end of a working process. Workflow automation aims the performance shift from humans to software programs for those activities. By automating workflows, organizations decrease the need for manual work and repetitive tasks. As a result, processes become more efficient as well as easier to track and monitor. Workflow automation is commonly extended by process orchestration [40].

Singular workflow steps comprise complex tasks such as data generation, translation and analysis, performance evaluation, and system-level verification among many others. Typically, operations in a workflow are dependent on the results from one or more of the preceding steps. Operations that share the same set of dependencies may be performed in parallel. Additionally, many workflows are iterative processes that are repeated until a predefined threshold is met. To capture these dependencies between operations, workflows are often represented as directed graphs [40].

Workflow automation platforms provide automation for singular processes within their respective domains. For the application presented here, performant interfaces between tools such as data and model transformations, passing of artifacts from one tool to another, and configuration of tools before execution are necessary. Workflow specification is based on generic activity graphs where each activity may involve the use of one or more domain-specific tools. The executable semantics runs in an execution engine. It, known as the workflow executor, can automate and monitor the execution of these processes [40].

Complex workflow which includes both data exchange with external partners and heterogeneous process orchestration presupposes some additional constraints to be addressed [41]:

- *Security*: Public networks are open to everybody; sensitive information needs to be exchanged securely.
- *Reliability*: Public networks are often not as stable as required, especially for transmission of large amounts of information (e.g. point cloud model file package which contains terabytes of data).
- *Traceability*: For the exchange in a globalized economic environment often—even legally binding—proof of execution, data transmission, and reception is required. Likewise, a way to check the intermediate results in a neutral format is necessary [42].
- *Efficiency*: Process robustness and stability (such as repeatable exchanges and defined content) without loss of competitiveness becomes more and more crucial.
- *Modularity*: Due to the different origins of singular software components, a general, modular structure is necessary e.g., by using process templates [43]. The flexible handling of the object recognition framework with the option of replacement by another must also be foreseen.

A general deployment schema for the generation of a digital twin with all optional services is depicted in Fig. 7.7 as a collaborative procedure among several parties. The acquisition of the plant in question or a portion of is conducted by the capture partner on the customer's premises. Then point cloud data are sent to OpenDESC.com where object recognition is done as described in Sect. 7.4. As an option, the preliminary results in neutral representation can be transmitted to the cooperation partner who provides the link with P&ID, imports the neutral representation of the plant in a proprietary environment (e.g., AVEVA) (please refer to Sect. 6.4), and finalizes the digital twin [38]. The role of the main contractor is defined case-by-case depending on the extent of work.

After the full data model of the plant was created, it is transmitted to the customer (e.g., EPC contractor) which completes the required change of a unit or the plant. Finally, the data set is exported to the process planning department in accordance with the extent of the planned work [43]. External data exchange is performed under high-level encryption. Workflow automation also includes the handling of alternative

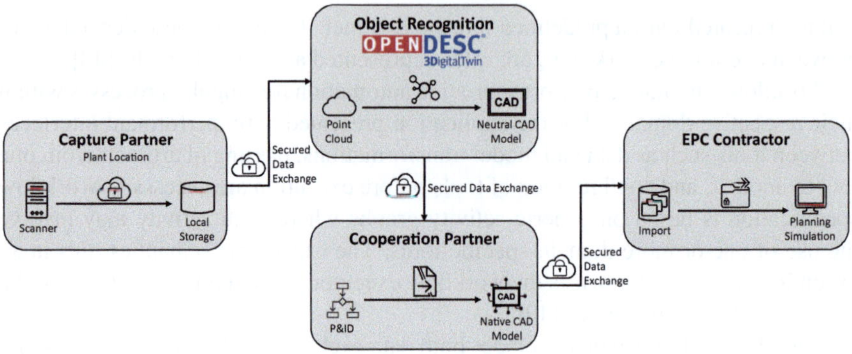

Fig. 7.7 Collaborative generation of digital twin [41]

routes and loops which are necessary to execute all the options. Customer-specific setup can be stored and reused.

7.6 Discussion and Future Perspectives

As presented in this chapter, a productive implementation of object recognition in process plants is supported by both the intensive research in the field of computer vision and practical applications which were deployed in wide areas of industry and society. Introduced by a large development community across the globe, a plethora of well-documented approaches, primarily based on supervised learning with CNN, are described in the literature and also available on software-sharing platforms [12, 17, 22–25, 29, 30, 35, 44–50]. Subsequently, dozens of review papers highlight the achievements in object recognition with different methods (image-based, point cloud based, and video-based) in different fields of application pointing out specific strengths and weaknesses of analyzed approaches [6, 8, 9, 13, 15, 16]. From this point of view, the current challenge is more to select and adapt one or more appropriate frameworks for object recognition that can be used and maintained in a mid-term period rather than to develop a new one.

The challenge for the generation of the digital twin of a process plant lies rather in the reconstruction of the structure of pipe systems in the plant than the reconstruction of building structures. In order to achieve a 3D reconstruction from scanned data the most common approach is applied here. Such approaches are appropriate for industrial sites and mechanical designs because most objects are built-up by prismatic elements. However, such bottom-up approaches heavily depend on locality and can hardly completely rebuild models such as plants with sufficient connectivity. Bottom-up rebuild of assemblies which comprise a multitude of small components is also sensitive to noise and outliers in the input data due to missing global and content-aware relations [23].

Pipelines are dominant constituents in process plants due to their functional importance and widespread presence. Therefore, the recognition of piping systems needs the recognition of their structures first. Piping systems consist of forked structures defined by long cylinders organized in dense and complex configurations with partial occlusion. Although pipes are mostly cylinders which can be easily defined by their axis and radius, a piping system includes further components such as flanges, valves, inlets, elbows, and tees which must be properly recognized before the piping system can be rebuilt [48]. Keeping in mind that 3D scanning and rebuilding of pipelines is error-prone due to several causes of error (discontinuity of scan, small pipe surfaces, intricate structure, self-occlusions, and occlusions by insulation), a fully automated process cannot be expected at this time.

Therefore, a highly automated, stable, and robust method for the recognition of piping systems from raw 3D scans is explored here. It combines the local features (pipe segments, fittings) with global properties (piping structure) to build structures (piping system segment). For all three building blocks, supervised training with CNN frameworks is applied. The robustness and stability of these methods can be further improved by additional training sets.

7.7 Conclusions and Outlook

This chapter presented the way how the digital twin of a process plant is implemented as a solution. The application of object recognition methods based on supervised learning is justified in terms of quality, robustness, and stability. A brief overview of methods for object recognition is presented with their classification according to input in image, point cloud, and video-based methods. Selection of a suitable method for practical implementation is difficult because essential parameters differ greatly for individual approaches. Otherwise, oversupply and too little differentiation of results can be observed there. The decision to point cloud from a laser scanner is justified as the preferred data source.

Despite their advantages in terms of accuracy and quality, the point cloud-based methods have a large potential in two directions [51]. While the effort to acquire a scan of a process plant makes still a significant share of the total costs, it should be further optimized by the utilization of more efficient scan devices. This improvement could yield a faster scan process as well as a more homogeneous output point cloud and, thus, reduce the need for checks by users [26]. This would also reduce the share of interactive work (Sect. 7.5.1).

Implementation of an attention mechanism could pose an alternative approach to identifying the content of interest in a huge environment and enforce the process of recognition. The attention mechanism greatly improves the efficiency and accuracy of perceptual information processing. Subsequently, attention mechanisms can be used as a resource allocation scheme, which is the main means to solve the problem of information overload. In the case of limited computing power, it can process more important information with limited computing resources [52].

While the recognition systems perform well in controlled environments, they are unable to generalize in less controlled environments. Therefore, an appropriate test base that represents the real conditions is necessary. Flexible input for test data from the open-access test data sets of objects of interest must be foreseen. Conversely, the most important objects should be further refined in distinctive parameters (dimensions, insulation, surface condition). The reduction of the training period and expenses will remain a requirement with a high priority [23].

In practical applications, CNNs struggle to generalize under altering scenarios, like recognizing the variability and heterogeneity of the instances of elements belonging to the same category. That is exactly the situation that can be observed in piping systems where singular objects are often distinct from each other by some detail features or objects with an identical function distinguish by secondary detail features [53].

Requirements from the software engineering point of view must also be taken into account: supporting the integration of code and documentation generation as well as explicitly linking knowledge base structure and meaningful content with the recognition application elements and code. When considering the introduction and use of object recognition applications and templates, the adaptability and maintainability of developed applications and templates remain a point of concern in these turbulent times. Templates should cover the entire process as well as the most expensive steps (segmentation, classification, determination of centerlines) [42]. Furthermore, the trend to real-time execution will continue and it will have a dominant impact on the software architecture [54].

References

1. Stjepandic J, Sommer M, Denkena B (2022) DigiTwin: an approach for production process optimization in a built environment. Springer International Publishing Switzerland. https://doi.org/10.1007/978-3-030-77539-1
2. Zhong B, Wu H, Ding L, Love PED, Li H, Luo H, Jiao L (2019) Mapping computer vision research in construction: developments, knowledge gaps and implications for research. Automation in Construction 107:102919. https://doi.org/10.1016/j.autcon.2019.102919.
3. Li F, Kim MK (2021) Mirror-aided registration-free geometric quality inspection of planar-type prefabricated elements using terrestrial laser scanning. Automation in Construction 121:103442. https://doi.org/10.1016/j.autcon.2020.103442
4. Kitaguchi D, Takeshita N, Hasegawa H, Ito M (2022) Artificial intelligence-based computer vision in surgery: recent advances and future perspectives. Annals of Gastroenterological Surgery 6:29–36. https://doi.org/10.1002/ags3.12513
5. Galán JJ, Carrasco RA, LaTorre A (2022) Military applications of machine learning: a bibliometric perspective. Mathematics 10:1397. https://doi.org/10.3390/math10091397
6. Chai J, Zeng H, Li A, Ngai EWT (2021) Deep learning in computer vision: a critical review of emerging techniques and application scenarios. Machine Learning with Applications 6:100134. https://doi.org/10.1016/j.mlwa.2021.100134
7. Shah SAA (2016) Surface representations and automatic feature learning for 3D object recognition. PhD thesis, University of Western Australia

8. Rani S, Lakhwani K, Kumar S (2022) Three dimensional objects recognition & pattern recognition technique; related challenges: a review. Multimedia Tools and Applications 81:17303–17346. https://doi.org/10.1007/s11042-022-12412-2
9. Xu S, Wang J, Shou W, Ngo T, Sadick AM, Wang X (2021) Computer vision techniques in construction: a critical review. Archives of Computational Methods in Engineering 28:3383–3397. https://doi.org/10.1007/s11831-020-09504-3
10. Sommer M, Seiffert K (2021) Scan methods and tools for reconstruction of built environments as basis for digital twins. In: Stjepandić J et al (eds) DigiTwin: an approach for production process optimization in a built environment. Springer Nature, pp 51–78. https://doi.org/10.1007/978-3-030-77539-1_4
11. Gong Z, Wen C, Wang C, Li J (2018) A target-free automatic self-calibration approach for multibeam laser scanners. IEEE Trans Instrum Meas 67(1):238–240. https://doi.org/10.1109/TIM.2017.2757148
12. Czerniawski T, Nahangi M, Haas C, Walbridge S (2016) Pipe spool recognition in cluttered point clouds using a curvature-based shape descriptor. Autom Constr 71:346–358. https://doi.org/10.1016/j.autcon.2016.08.011
13. Salem B, Stjepandić J, Stobrawa S (2019) Assessment of methods for industrial indoor object recognition. Advances in Transdisciplinary Engineering 10:390–399. https://doi.org/10.3233/ATDE190145
14. Riveiro B, Lindenbergh R (eds) (2020) Laser scanning: an emerging technology in structural engineering. CRC Press/Balkema, Leiden. https://doi.org/10.1201/9781351018869
15. Kaur R, Singh S (2022) A comprehensive review of object detection with deep learning. Digital Signal Processing 132:103812. https://doi.org/10.1016/j.dsp.2022.103812
16. Qian R, Lai X, Li X (2022) 3D Object detection for autonomous driving: a survey. Pattern Recognition 130:108796. https://doi.org/10.1016/j.patcog.2022.108796
17. Wang K, Zhang Z, Wu X, Zhang L (2022) Multi-class object detection in tunnels from 3D point clouds: an auto-optimized lazy learning approach. Advanced Engineering Informatics 52:101543. https://doi.org/10.1016/j.aei.2022.101543
18. Weinmann M (2020) Semantic segmentation of dense point clouds. In: Riveiro B, Lindenbergh R (eds) Laser scanning: an emerging technology in structural engineering. CRC Press/Balkema, Leiden, pp 93–111. https://doi.org/10.1201/9781351018869
19. Poza-Lujan J-L, Posadas-Yagüe J-L, Simó-Ten J-E, Blanes F (2020) Distributed architecture to integrate sensor information: object recognition for smart cities. Sensors 20:112. https://doi.org/10.3390/s20010112
20. Stjepandić J, Sommer M (2022) Object recognition methods in a built environment. In: Stjepandić J, Sommer M, Denkena B (eds) DigiTwin: an approach for production process optimization in a built environment. Springer series in advanced manufacturing. Springer, Cham, pp 103–134. https://doi.org/10.1007/978-3-030-77539-1_6
21. Symeonidis G, Nerantzis E, Kazakis A, Papakostas GA (2022) MLOps—definitions, tools and challenges. In: 2022 IEEE 12th annual computing and communication workshop and conference (CCWC), Las Vegas, NV, USA, pp 0453–0460. https://doi.org/10.1109/CCWC54503.2022.9720902
22. Song Y, He F, Fan L, Dai J, Guo Q (2022) DSACNN: dynamically local self-attention CNN for 3D point cloud analysis. Advanced Engineering Informatics 54:101803. https://doi.org/10.1016/j.aei.2022.101803
23. Cheng L, Wei Z, Sun M, Xin S, Sharf A, Li Y, Chen B, Tu C (2020) DeepPipes: learning 3D pipelines reconstruction from point clouds. Graph Models 111:101079. https://doi.org/10.1016/j.gmod.2020.101079
24. Agapaki E, Brilakis I (2020) CLOI-NET: class segmentation of industrial facilities' point cloud datasets. Advanced Engineering Informatics 45:101121. https://doi.org/10.1016/j.aei.2020.101121
25. Romero-Jarén R, Arranz JJ (2021) Automatic segmentation and classification of BIM elements from point clouds. Autom Constr 124:103576. https://doi.org/10.1016/j.autcon.2021.103576

26. Verykokou S, Ioannidis C (2023) An overview on image-based and scanner-based 3D modeling technologies. Sensors 23:596. https://doi.org/10.3390/s23020596
27. Monsone C, Csapó ÁB (2023) Towards rapid prototyping of digital twins based on hand-held video. In: Klempous R, Nikodem J, Baranyi PZ (eds) Accentuated innovations in cognitive info-communication. Topics in intelligent engineering and informatics, vol 16. Springer, Cham, pp 167–181. https://doi.org/10.1007/978-3-031-10956-0_8
28. Singh RD, Mittal A, Bhatia RK (2019) 3D convolutional neural network for object recognition: a review. Multimedia Tools and Applications 78:15951–15995. https://doi.org/10.1007/s11042-018-6912-6
29. Rangel JC, Martínez-Gómez J, Romero-González C, García-Varea I, Cazorla M (2018) Semi-supervised 3D object recognition through CNN labeling. Applied Soft Computing Journal 65:603–613. https://doi.org/10.1016/j.asoc.2018.02.005
30. Kousik NV, Natarajan Y, Raja RA, Kallam S, Patan R, Gandomi AH (2021) Improved salient object detection using hybrid Convolution Recurrent Neural Network. Expert Systems with Applications 166:114064. https://doi.org/10.1016/j.eswa.2020.114064
31. Abbas Zaidi SS, Ansari MS, Aslam A, Kanwal N, Asghar M, Lee B (2022) A survey of modern deep learning based object detection models. Digital Signal Processing 126:103514. https://doi.org/10.1016/j.dsp.2022.103514
32. Zhang Y, Wang L, Dai Y (2023) PLOT: a 3D point cloud object detection network for autonomous driving. Robotica 1–17. S0263574722001837. https://doi.org/10.1017/S0263574722001837
33. Hafiz AM, Hassaballah M, Binbusayyis A (2023) Formula-driven supervised learning in computer vision: a literature survey. Appl Sci 13(2):723. https://doi.org/10.3390/app13020723
34. Azeem M, Haleem A, Javaid M (2022) Symbiotic relationship between machine learning and industry 4.0: a review. J Ind Intg Manag 7(3):401–433. https://doi.org/10.1142/S2424862221300027
35. Pratama ARP, Dewantara BSB, Sari DM, Pramadihanto D (2022) Density-based clustering for 3D stacked pipe object recognition using directly-given point cloud data on convolutional neural network. EMITTER International Journal of Engineering Technology 10(1):153–169. https://doi.org/10.24003/emitter.v10i1.704
36. Stjepandić J, Sommer M (2022) Object recognition findings in a built environment. In: Stjepandić J, Sommer M, Denkena B (eds) DigiTwin: an approach for production process optimization in a built environment. Springer series in advanced manufacturing. Springer, Cham, pp 155–180. https://doi.org/10.1007/978-3-030-77539-1_8
37. Grau M, Korol W, Lützenberger J, Stjepandić J (2021) Automated generation of a digital twin of a process plant by using 3D scan and artificial intelligence. Advances in Transdisciplinary Engineering 16:93–102. https://doi.org/10.3233/ATDE210087
38. Kremer P, Lützenberger J, Müller F, Stjepandić J (2022) An approach for the incremental update of a digital twin of a process plant. Advances in Transdisciplinary Engineering 28:310–319. https://doi.org/10.3233/ATDE220660
39. Zhang S, Huang K, Xu Z (2022) Re-thinking model robustness from stability: a new insight to defend adversarial examples. Mach Learn 111:2489–2513. https://doi.org/10.1007/s10994-022-06186-9
40. Hartsell C, Mahadevan N, Nine H, Bapty T, Dubey A, Karsai G (2020) Workflow automation for cyber physical system development processes. In: 2020 IEEE workshop on design automation for CPS and IoT (DESTION), Sydney, NSW, Australia, pp 1–9. https://doi.org/10.1109/DESTION50928.2020.00007
41. www.opendesc.com. Accessed 1 March 2023
42. Pfouga A, Stjepandić J (2018) Leveraging 3D geometric knowledge in the product lifecycle based on industrial standards. Journal of Computational Design and Engineering 5(1):54–67. https://doi.org/10.1016/j.jcde.2017.11.002
43. Kuhn O, Liese H, Stjepandić J (2011) Methodology for knowledge-based engineering template update. In: Cavallucci D, Guio R, Cascini G (eds) Building innovation pipelines through computer-aided innovation. Springer, Berlin, pp 178–191. https://doi.org/10.1007/978-3-642-22182-8_14

44. Moritani R, Kanai S, Date H, Watanabe M, Nakano T, Yamauchi Y (2019) Cylinder-based efficient and robust registration and model fitting of laser-scanned point clouds for as-built modeling of piping systems. Computer-Aided Design Applications 16(3):396–412. https://doi.org/10.14733/cadaps.2019.396-412
45. Araújo AMC, Oliveira MM (2020) Connectivity-based cylinder detection in unorganized point clouds. Pattern Recognition 100:107161. https://doi.org/10.1016/j.patcog.2019.107161
46. Bergamasco F, Pistellato M, Albarelli A, Torsello A (2020) Cylinders extraction in non-oriented point clouds as a clustering problem. Pattern Recognition 107:107443. https://doi.org/10.1016/j.patcog.2020.107443
47. Wang B, Yin C, Luo H, Cheng JCP, Wang Q (2021) Fully automated generation of parametric BIM for MEP scenes based on terrestrial laser scanning data. Automation in Construction 125:103615. https://doi.org/10.1016/j.autcon.2021.103615
48. Lee ID, Lee I, Han S (2021) 3D reconstruction of as-built model of plant piping system from point clouds and port information. Journal of Computational Design and Engineering 8(1):195–209. https://doi.org/10.1093/jcde/qwaa072
49. Oh I, Ko KH (2021) Automated recognition of 3D pipelines from point clouds. Vis Comput 37:1385–1400. https://doi.org/10.1007/s00371-020-01872-y
50. Xie Y, Li S, Liu T, Cai Y (2023) As-built BIM reconstruction of piping systems using PipeNet. Automation in Construction 147:104735. https://doi.org/10.1016/j.autcon.2022.104735
51. Wang Q, Li J, Tang X, Zhang X (2022) How data quality affects model quality in scan-to-BIM: a case study of MEP scenes. Automation in Construction 144:104598. https://doi.org/10.1016/j.autcon.2022.104598
52. Niu Z, Zhong G, Yu H (2021) A review on the attention mechanism of deep learning. Neurocomputing 452:48–62. https://doi.org/10.1016/j.neucom.2021.03.091
53. Sommer M, Stjepandić J, Stobrawa S, von Soden M (2023) Automated generation of digital twin for a built environment using scan and object detection as input for production planning. Journal of Industrial Information Integration 33:100462. https://doi.org/10.1016/j.jii.2023.100462
54. Ghasemi Y, Jeong H, Choi SH, Park KB, Lee JY (2022) Deep learning-based object detection in augmented reality: a systematic review. Computers in Industry 139:103661. https://doi.org/10.1016/j.compind.2022.103661

Chapter 8
Practical Application of Digital Twin of a Process Plant

Abstract The digitalization of a brownfield process plant and the generation of a digital twin respectively is a time-consuming and expensive task. The as-is state needs to be captured and engineers need to remodel the plant manually within the CAD system. By applying object recognition from point cloud this task can be automated. However, various factors influence the results. Object recognition from a point cloud highly depends on the quality of a point cloud. Environmental influences such as dust, dirt, vapors, and more are captured as well and need to be filtered out. Additionally, according to the specific domain also constructive influences appear. Depending on the size of the plants the extent of used steel construction or additional platforms and stairs or ladders can be found. These elements hurdle the scanning activities and lead to scanning shadows. In consequence, fragmentations appear, which reduce the quality of the results. Here also the density regarding the packaging of the piping system and supporting constructive structures play a role. Domain-specific attributes of piping systems need to be considered as well. For some domains, the use of highly polished materials is required while others are made for more rough environments. Additional aspects such as insulation or painting change the appearance to some extent. All these factors affect the object recognition process as well. In addition, components' specific properties come into place. While some components are simple connections between two components, others lead to a change of direction within the course of a piping system or even change attributes such as diameter. All that makes post-processing and optimization of parameters necessary. The practical application is presented in the context of examples from different domains. Various point clouds are analyzed according to the above-mentioned factors. Finally, the automatically generated piping systems are presented and approved in terms of integration with the original point cloud.

Keywords Digital Twin · Product Lifecycle Management · Object recognition from point cloud · Process industry · Post-processing of piping systems

J. Stjepandić et al., *Generation and Update of a Digital Twin in a Process Plant*,
https://doi.org/10.1007/978-3-031-47316-6_8

8.1 Introduction

The conceptual implementation presented in Chap. 7 has been applied to plants from various branches. While the circumstances of their operations vary the components they are composed of are standardized by standardization bodies such as DIN, ISO, EN, etc. [1]. They define specific component versions and their executions about the material, nominal diameter, the thickness of walls, length, angles, etc. [2]. This way the engineers planning such a plant can rely on standardized components about pipes, elbows, tee, etc. [3].

During a planning project frame conditions need to be considered. These can be dedicated to the branch or the specific application [4]. Generally spoken, e.g. specific materials need to be chosen according to the media inside the pipes. This way, chemical reactions between the media and the material of the pipes can be avoided. E.g. in some cases, a chemical reaction between the media and the pipe can lead to corrosion of the pipes. This has effects on various levels. Firstly, corrosion reduces the security/safety of a pipe. A secure operation of a plant is probably one of the most important situations operators are willing to achieve [5]. By choosing the best material in accordance with the media inside the pipe, such chemical reactions can be avoided [6].

But not only effects on the pipeline side are necessary to be avoided. Also, chemical reactions that lead to a change in the state of the media inside the pipes need to be avoided. Such reactions can change the chemical composition of the media and this way can have direct effects on the product quality.

Dedicated to these branches specific requirements regarding hygiene need to be fulfilled. In the food industry or the pharmaceutical sector e.g. the requirements are extremely high. Specific materials that fulfill these requirements need to be chosen here.

Design engineers in the field of process industry apply for their work in CAD systems which provide component catalogs [7]. These catalogs provide the standard components defined by norms such as DIN, ISO, EN, etc. [8]. Even though the designer only needs to make the right choice of components to compose the piping systems, the number of available components is so high, that uncertainties in the choice of components are too high. To reduce these uncertainties and to support the engineers with the right choice, so-called pipe specifications are defined. These specifications or piping classes are subsets of the applied norms. They include a selection of components that are available for the planning of a plant project.

The specification offers a bunch of components that can be chosen by the engineers during the planning process. This way it can be guaranteed, that only components can be chosen that fit together. The nominal diameter of the components here is the leading attribute. However, as mentioned above further properties come into place to define components.

When a plant has been constructed and reached the operations phase it undergoes ongoing maintenance and repair activities [9]. This leads to undocumented changes in the state of a plant. Additionally, during the construction of a plant, the composition varies in comparison to the as-designed state. This way, e.g. the length of pipes might vary and further connection points, such as flanges, are shifted slightly in comparison to the as-designed state. For that reason, when it comes to modernization activities, the digitization of the as-is state is the first step [10]. The current physical state needs to be transferred into a 3D digital representation [11]. Therefore, various approaches can be carried out (see Chap. 7).

3D laser scanning is the latest and most common methodology for capturing and digitizing plants within one process step. A laser scans the environment and captures a point when the laser hits a surface. This way a point cloud is generated representing the surfaces of an asset in its current state [12].

Even though high-resolution laser scanning systems provide technology enabling capturing with high quality and accuracy, they also capture dirt, dust, vapors, reflections, etc. These uncertainties regarding the exact shapes of piping components hurdle the object recognition process. Nevertheless, the presented approach in Chap. 6 enables object recognition of piping components in a two-step process based on artificial intelligence [13].

The most critical task, when it comes to object recognition is the identification of the various components a piping system is composed of. Even though the components themselves are standardized as mentioned above, the number of possible combinations is endless [14].

In this chapter, the complexity with regard to different types of components as well as the specialties of the various branches is provided. The major challenge is to understand the hurdles from an IT perspective to achieve reliable recognition and provide an outcome that suits the client's expectations from an engineering perspective [15].

Therefore, Sect. 8.2 focuses on the application at a component level. Section 8.3 takes specialties concerning the different industries into account. Various projects already carried out are presented as samples. The generation of the CAD model and its proof is provided in Sect. 8.4. Finally, the results are discussed and an outlook is provided in Sect. 8.5.

8.2 Specific Properties of Components

The scan of a plant covers all components and objects existent in a plant. During the object recognition process, the overall structure of a plant is broken down into major areas (equipment, structure, and piping). This way the complexity is reduced and the different parts are differentiated according to their functionality. Hereby the focus relies on the piping system. Remodeling the piping system is the most time-consuming task in comparison to arranging the equipment in space or remodeling of steel structure [16].

Even though the focus relies on the piping system, it is still of enormous complexity. The entire system is composed of several sub-systems called pipe strings that connect equipment (see Sect. 6.3). Pipe springs connect two or more equipment units. In some cases, there is a direct connection between two equipment units. The 3D expression is not reflected by this definition. Pipe strings can be short connections with a minimum of components as well as long-ranging complex courses in space. In the case of more than two equipment units being connected, pipe strings are differentiated into branches. These branches can be connected by components called tees or even crosses. This way a complex system of pipe strings can be established [3].

All pipe strings or branches respectively are composed of standard parts. These standard parts have differences with regard to their appearance according to their use. While straight pipes reflect a straight connection between two components in a system, an elbow allows a change of direction. While the mentioned example shows the use in the context of the course of a piping system also components with regards to change of properties are available. In this sense, reducers enable the engineers to connect to pipes with different diameters. This might become necessary if the fluid speed needs to be increased by a reduction of the nominal diameter. As already mentioned also components for separation and merge are needed. Through tees or crosses a linkage of various branches can be applied. In addition, components can be connected by applying flanges, e.g. by connecting a valve within a pipe string. Flanges are also needed to connect the piping system to the equipment. The flanges on the equipment side are called nozzles [3].

All mentioned physical properties are dedicated to an engineering perspective. However, for the generation of a digital twin, these components need to be analyzed from an IT and object recognition perspective [17]. Table 8.1 provides an overview of the mentioned components with their specific properties and the challenges appearing with regard to the recognition process.

Pipe—A pipe is a standardized straight connection between two components. Hereby the connection can be between two other components (e.g. flange-pipe-elbow), between the same components (e.g. pipe-pipe-pipe), or a combination of the same and other components (e.g. pipe-pipe-elbow). In addition, pipes can be insulated to protect the pipes from temperature deviations. Two major challenges appear with regard to object recognition and deriving geometrical parameters. Firstly, straight pipes can be directly connected. This seamless connection harms the recognition on the instance level. Secondly, parts can be insulated. The insulation adds measurement to the original surface of a pipe. Since a scanner is only able to capture the surface and it is not possible to distinguish between insulated pipes and original pipes the accuracy in terms of metadata is significantly reduced [11].

Elbow—While pipes and reducers are straight connections between two components an elbow enables a change of direction without changing the nominal diameter. The direction can be changed according to standardized angles such as 90°, 60°, 45°, and further. Multiple elbows can be combined. Two major challenges appear in this context. The determination of the correct angle on the one hand. On the other hand, each elbow needs to be recognized in assemblies with multiple elbows directly connected.

Bend—Comparable with elbows also bends enable a change of direction in the course of a piping system. However, bends are pipes that are bent for individual

Table 8.1 Specific properties of piping components, derived from [2] and Fig. 6.2

Component	Specific property	Challenges
Pipe	Connection of two components, straight, insulated, or uninsulated, inner and outer diameter	Several pipes can be connected seamlessly, distinguishing between isolated or unisolated parts
Elbow	Change of direction (standardized angles), inner and outer diameter	Determination of angle, a combination of multiple elbows is possible
Bend	Change of direction (non-standardized angles), inner and outer diameter	Determination of angle, limitation against elbow
Reducer	Connection of two components, reduction of nominal diameter, Execution: straight or eccentric	Determination of diameters
Tee	Connection of two pipe strings, outgoing diameter usually smaller than the diameter of the main axis	Determination of diameters, determination of angles
Intersection	Connection of three pipe strings, various diameters possible	Determination of diameters, determination of angles
Flange	Connection between piping and armatures, instruments or equipment,	Differentiation of flange connections, consideration of sealing
Valve	Connection between two components, regulation of media flow	Determination of nominal diameter
Nozzle	Connection of the point of equipment to the piping system	Differentiation of flange connections, consideration of sealing

purposes. They are not standardized and the angles can vary. The determination of the angle and differentiation to standardized elbows are the major goals to achieve.

Reducer—Analogue to the pipes also reducers connect two components with each other. A reducer changes the nominal diameter and can be executed in a straight or eccentric manner. This way components with different diameters can be connected. As pipes, they can be insulated to protect the media from temperature changes. With regard to geometrical properties, the diameters of both ends need to be determined correctly.

Tee—As mentioned above, a pipe string can be composed of sub-systems called branches. These branches can be connected by tees. These components are standardized and connected to branches at a defined angle. Hereby, one branch is reflected by the main axis and the second branch by the outgoing axis. The diameter of the outgoing axis can be smaller in comparison to the main axis. Even though tees are standardized parts, the angle between both axes may vary. Determining the correct angle as well as the correct diameters of both axes are the challenges here.

Intersection—A pipe string composed of three branches demands an additional linkage in comparison to a tee. Therefore, an intersection is implemented. Intersections are standardized components. Again, the axis of the outgoing axis can be smaller than the diameter of the main axis. Angles may vary. For that reason, also here the determination of the diameters as well as the correct angles are the major challenges as well.

Flange—A flange is a standardized part enabling the connection of two components. In this sense, e.g., armatures or instruments can be attached to a pipe string. Further, the pipe string can be connected to the equipment. Flanges enable a connection without welding and an easy change of components respectively. To ensure the tightness of the flange connection are sealed with sealings. Distinguishing between the various components of a flange connection is the big hurdle here.

Valve—A valve is implemented within a pipe string for the regulation of the media flow inside the pipes. In the context of maintenance and repair activities, these components should be replaced easily. Therefore, valves are often equipped with flanges. This way they can be integrated into the piping system by flange connections. For the control of the valves, these are equipped with hand wheels or levers. The major challenge is the determination of the nominal diameter.

Nozzle—Nozzles are standardized connection points on the equipment side. They can be seen as flanges welded to equipment. They are the connection point for piping systems. Their standardization is according to the standardization of flanges, to ensure a connection. Usually, nozzles are connected to flanges and tightened by sealings between the flange and nozzle. This way a flange nozzle connection can be considered in the same way as a flange connection. For that reason also here the major challenge is the differentiation between the components.

The mentioned components represent the typical components pipe systems are composed of. Of course, there is no claim of completeness. Additional component types are available in the market and used. Additionally, specific parts or systems of specialized companies, as well as home-grown developments, are not considered in this list.

8.3 Complexity of Plants

Object recognition in the field of process plants is a challenging task. Even though the utmost piping components are standardized and therefore precisely defined the determination of them is a challenge. In Sect. 8.2 typical components have been analyzed according to their properties and the challenges from an IT perspective for a sufficient generation of a digital twin. In addition, further challenges appear.

Process plants are applied in various domains such as chemical, pharmaceutical, refinery, energy, and further [18]. And each domain comes with specific properties concerning the environment. While refineries can be widely spread throughout several square kilometers a biogas plant is pretty small. However, both have large parts of their piping system outside. Humidity, dust, and dirt from the surrounding nature are around and can be attached to the piping system. This stands in big contrast to plants from the pharmaceutical industry. Here hygienic is a critical factor. To ensure sanitation throughout the entire plant, other materials for the piping system are used (e.g. stainless steel) and they are implemented in buildings. This leads to various challenges during capturing the as-is state with devices. While in a refinery the pipes may be attached with dirt and depending on the weather conditions fog or dust needs to be expected, in the pharmaceutical domain the extreme cleanliness and the reflections from the polished stainless steel pipes are the challenge. These examples illustrate the enormous influence of environmental conditions [19]. The quality of the scan is expected to vary a lot as the challenges regarding object recognition respectively.

Besides the environment also further factors influence object recognition. One major factor is the complexity of the individual process plant [20]. As samples may serve a biogas plant on the one hand and a big chemical plant on the other hand. Both consist of piping systems made up of pipe strings, which again are composed of standard parts. However, a biogas plant is comparably small and comes along with small complexity. Most of the technical parts can be implemented within an overseas container. Additional pipe strings connect the equipment. The amount of process stages is low and no bigger auxiliary systems are needed to operate a biogas plant [21].

Within a chemical plant, raw material is processed throughout several reaction stages (see Sect. 4.2.1) to a final product. To enable this process auxiliary processes are needed to produce intermediate products or support the main process with energy, etc. Each auxiliary process is enabled by an additional piping system. Since all processes and piping systems respectively are connected, the complexity increases significantly in comparison to the biogas plant.

Also, the density of the pipes is a factor regarding the quality of a scan. The higher the density of the pipes, the more effort is needed to achieve a comprehensive point cloud and ensure reduced fragmentation in terms of scan shadows. Fragmentation lowers the scan quality concerning the reliability of the object recognition process [12].

Even if a comprehensive scan is achieved and interfering factors are minimalized the physical state of the scanned object influences the sufficiency of the object recognition process [22]. During the planning CAD systems are used. Ideal components are assembled. In the real world, parts are occupied with inaccuracies. Standardized angles are not perfectly met, and pipes are not perfectly straight or cylindric, to name just a few. Over time due to gravity, heavy parts will change their shape to some extent. Long pipes with a dead weight bend through over time.

Depending on the size of a plant, the number of necessary scans varies. Big plants or plants with a high density of pipes make more scans necessary. These need to be registered with each other to ensure that the single point clouds are merged into a comprehensive point cloud. In this sense, the size of a plant reflects the number of necessary scans for the complete capturing of a plant [12].

Figure 8.1 shows an overview of typical plants from various domains in the process industry. Within the domains, an estimated average is taken to achieve positioning within the matrix. Deviations within a domain towards the different dimensions are possible. The matrix aims to achieve an overview of plants regarding their possible recognition.

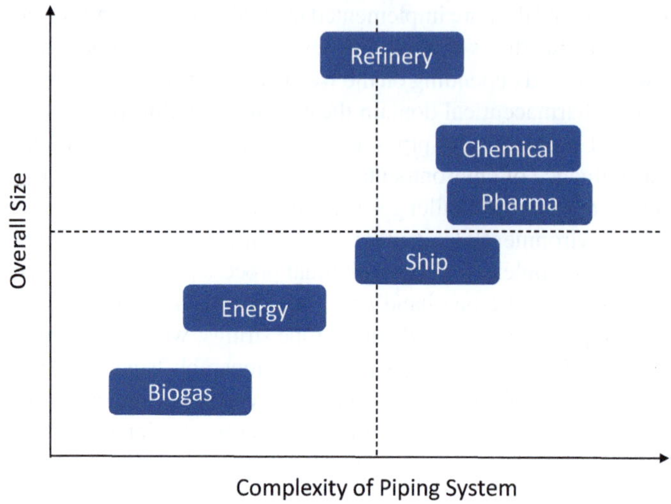

Fig. 8.1 Size-complexity map of industrial plants

Biogas plant—The biogas plant is mentioned above a small size plant. The capturing of the as-is state is achieved with comparably minor effort. Most of the pipe strings are arranged with low density, which enables better scanning with less fragmentation.

Energy plant—Plants in this domain may vary in terms of size and complexity. However, the processes applied are widely known and proven. Different systems such as piping, and HVACs are integrated but with regards to a simple product (energy in terms of heat or power). However, the number of reaction stages is low, and the number of auxiliary processes respectively low as well. Size and complexity are low.

Ship—Ships are not dedicated to the branch of the process industry. Nevertheless, various piping systems are implemented in ships. Taking cruise liners as a sample, this kind of ship contains fresh, grey, and black water systems, piping systems dedicated to the machinery of the ship, huge laundries, HVAC systems, and further. They reach throughout the different decks and departments. Therefore, the complexity is higher in comparison to other piping systems in the process industry. Even though cruise liners can extend to a length of 400 m, the size is limited and therefore the size can be seen as comparably average.

Chemical—The chemical industry is a wide field in terms of products that can be dedicated to this domain. Therefore, an average estimation is taken to find an appropriate position in the matrix. Deviations in all dimensions are possible. In a chemical process plant a raw material is processed through various reaction stages to a final product. Since a huge amount of material is produced, the size of such plants can be seen as above average. Auxiliary systems e.g., providing the main process with intermediate products, reaction material or energy leads to a linkage of many different systems. For that reason, the complexity of such plants is high.

Pharmaceutical—Within the pharma domain special requirements are raised according to sanitation. Therefore, subsystems can be capsulated to avoid any contradiction with other materials. In this sense also the piping system is made of material that does not react with other materials—stainless steel. In addition, the surfaces are polished. In comparison to other domains in the process industry, pharmaceutical plants are in average with regards to size a bit below average. However, due to the size of the pipes and the density of the piping system, the complexity is high.

Refinery—These plants can have extensions of several square kilometers. Therefore, they are the domain with the biggest sized plants. Even though the general processes are known and the complexity can be seen as average in comparison to other domains. Depending on the products produced, other plants do need much more auxiliary systems to facilitate a sufficient process.

The mentioned examples provide an overview of expectable process plant projects. Of course, individual plants may deviate from the overview. Estimations of averages have been taken to achieve this overview. Some of the mentioned domains have been exemplarily analyzed in more detail. The results are presented in the following sections.

8.3.1 Biogas Plant

A biogas plant is a process plant, that converts biomass into energy. The biomass can range from slurry, manure, and further organic material of animal origin to organic materials from food production. Here it is distinguished between agricultural and non-agricultural biogas production. During the fermentation process, biogas is generated. This can be used in a combined heat and power plant to generate energy in terms of electricity or heat. Another option is the refinery of the biogas into biomethane. It can be directly fed into the existing gas infrastructure.

In comparison to plants from other sectors such as the chemical industry biogas plants are standardized products. Even though biogas plants also have multiple variants dedicated to the biomass it is fed with, all variants are standardized. They can be ordered and assembled without any additional design effort. Also, their size and complexity are low in comparison to other industrial sectors [21].

Figure 8.2 shows an excerpt of a point cloud of a biogas plant. The control systems are arranged within a 12 ft overseas container. In addition, also the piping system is partially installed into the container. On top of the container, a heat exchanger is placed. Also, HVACs and a chimney can be seen here. In front of the container, several small size tanks, as well as pump units, are placed. All mentioned sort of equipment is connected by the piping system. It consists of straight pipes, elbows, tees, valves, and reducers. They are fixed by the pipe support systems.

In the following section the provided scan is shown in Fig. 8.2 is analyzed from an IT perspective. It is evaluated with regard to object recognition.

The light conditions can vary since the system is implemented partially outside and partially inside the overseas container. For that reason, the same types of pipes

Fig. 8.2 Excerpt of the point cloud of a Biogas Plant

are captured under artificial and natural light, which affects the recognition process. However, due to the low density of the packing of the piping system, the surfaces can be captured very well from 360° in most cases. This leads to a high scan quality and further to a high object recognition rate.

As mentioned above the plant is partially placed inside a 12 ft overseas container. This illustrates the overall size of the plant. In comparison to plants in other branches, the size of a biogas plant is small. The object recognition process benefits from it, e.g., in terms of process time.

Also, the components installed vary for example with regard to different nominal diameters. However, the total variation of diameters can be seen as low in comparison to more complex plants. Additionally, also the variation of the installed equipment is low. Many equipment units are comparable such as the tanks. Further, standardized equipment such as heat exchangers can be found, which can also be seen as a kind of standard part. All of that enables reliable object recognition of components and the piping system respectively.

The scan of the biogas plant is comparably clean. Only less dust, vaporization, or dirt has been captured. Also, reflections or noise can only be found minorly. That leads to a clear capture of the shapes of the components, which can be recognized very well respectively.

8.3.2 Chemical Plant

Chemical plants transform raw materials into final products. To achieve this goal multiple reaction stages (chemical reactions) need to be processed. Within these stages, the material is changing state such as a combination of two or more materials/ elements, differentiation of materials, change of material properties, etc. This way raw materials can be transformed into a product.

To enable a chemical reaction between materials this needs to be prepared usually. The preparation is carried out within so-called auxiliary systems. In these systems mechanical treatment as well as chemical treatments are possible. Depending on the product, that is envisaged to be produced, the combination of the main process and auxiliary processes leads to the high complexity of such plants [23].

Figure 8.3 shows an excerpt of a scan of a chemical plant. Because of the location of the piping system in a production hall, it is illuminated by artificial light. The ground floor is made of concrete, while the different levels are made of steel grids. To enable the workers to reach the upper levels, they are connected by stairs, also made of steel grids. The stairs as well as the upper levels are secured by railings. Lamps on top shine through these grids, which leads to many shadows on the lower levels.

Mainly two types of equipment are located in the plant: huge equipment such as silos or tanks as well as smaller equipment such as pumps. The huge equipment units are silos exploited to store raw materials as well as intermediate material stages within the different production stages. Pumps deliver the material throughout the piping system which connects the different reaction stages and equipment. It is composed of pipe strings, that are mainly of small and medium-sized diameters. In addition, flexible pipe strings can be found. They are connected to the silos and are used to transfer media from barrels into silos. These barrels are placed on pallets in the ways of the ground floor. Besides the piping system also HVACs are part of the plant.

Fig. 8.3 Excerpt of the point cloud of a chemical plant

Subsequently, the sample of a chemical plant is analyzed from an IT and object recognition perspective. As mentioned above, the chemical plant is illuminated by artificial light. It is provided by lamps positioned at the top of each level. Since the levels, except the ground floor each level is made of grids, the light shines through these grids and causes shadows. The shadows look like grid patterns that appear on the silos or tanks. These shadows harm the scan quality and the object recognition respectively.

Further, piping strings, equipment, and steel construction are placed close to each other. This makes a 360° scan of all components difficult and in many cases impossible. A consequence of this is scan shadows. These lead to a fragmentation of the scan. Also, this reduces the quality of the scan and the object recognition process respectively.

The huge equipment is placed centrally. Their extension ranges throughout various levels. From a scanning perspective, it is difficult to achieve a continuous scan of the equipment. The grid floors of the upper levels divide the equipment into several parts in terms of a point cloud. Fragmentation of the point cloud or scan shadows as the consequence, which reduces the scan quality and the object recognition process as mentioned above.

In comparison to the before mentioned biogas plant the chemical plant does not only consist of rigid pipes. Also, flexible pipes are integrated into the piping system. While straight pipes can change their orientation in space, flexible pipes can also change their shape and angle according to how they have been used. Object recognition regarding flexible pipes is more complex and for that reason a challenge.

Besides the core components such as piping components or equipment also barrels are placed on the ground floor. Barrels have a cylindric shape such as piping components. The object recognition process needs to ensure that the barrels are recognized as objects, not piping components.

On the upper levels, railings secure workers from falling. These are cylindric parts with a diameter. Basically, comparable with straight rigid pipes. Even though the standardization of the railings can be considered during the object recognition process, it is necessary to ensure that the railings are not identified as individual piping systems.

8.3.3 Power Plant

Power plants have the purpose of transforming raw materials into energy in terms of electrical energy or heat. To achieve this goal a combination of equipment and a piping system is arranged. Depending on frame conditions such as available source material (gas, oil, etc.), expected power volume, and further power plants are designed [24].

Figure 8.4 provides an excerpt of a power plant from the north of Germany operated by a public body. Equipment such as tanks and pump units are installed. They are connected by piping systems. These piping systems are distinguished by their purpose, which is reflected in the different colors of the pipes (e.g., green or silver). To reduce loss of heat, some pipes are also insulated. All are supported by pipe supports fixed on the steel structure. The whole plant is an integrated product consisting of equipment, a piping system, and a steel structure. To reach the different areas, platforms are installed to enable the workers to carry out their work. Stairs are connecting the different platforms.

Fig. 8.4 Excerpt of the point cloud of a power plant

In the present example of a power plant (Fig. 8.4), it can be seen, that the piping system is placed inside a building. Therefore, it is illuminated by artificial light. This affects the appearance of the coloring of the various pipe strings.

To ensure continuity in the processes, some parts of the system are installed redundantly [25]. Therefore, equipment such as the green tanks on the ground floor is three times installed. They are connected by pipe strings with medium-sized diameters. Also, pumps can be found even though they are not visible in Fig. 8.4.

The various equipment is connected by the piping system that runs through the different levels. These levels are made of a grid. They are connected by stairs made of a grid as well. The artificial light shines through the grids with the result of shadows visible on the piping system and the equipment. This influences the coloring and the sufficiency of the recognition process respectively.

At the right part of the scan, many fragments can be found. This fragmentation is affected by scan shadows. Additional scanning would be necessary to achieve a consistent point cloud. The high degree of fragmentation reduces the scan quality significantly and hurdles the recognition process.

Piping supports connect the piping systems with the steel structure which also carries the construction of the building itself. The steel structure contradicts the scan quality in terms of scan shadows. Objects behind the steel structure are not well captured. In the right part of the point cloud, such fragmentation can be found. These inconsistencies are within the point cloud of the piping system causing gaps that need to be closed logically.

Besides the piping system of the main process, auxiliary piping systems can be found as well as HVACs. While HVACs often have a rectangular shape, here they have a cylindric design. Correct recognition and delamination of the piping system are challenging the recognition process.

8.3.4 Refinery

A refinery is a specialized type of chemical plant. It is dedicated to the transformation of raw oil into various products such as Diesel, gas, heating oil, etc. These types of plants are large-scale plants. They can reach square kilometers in terms of size and kilometers of piping system respectively. Accordingly, the complexity of such plants is very high and the number of different piping components as well as equipment is enormous [26].

As shown in Fig. 8.5 the piping system is composed of pipes with different diameters. Pipes with and without insulation can be found. Also, the variation with regard to the diameter is high as well. The pipe string connects the different equipment. However, due to redundant systems and huge distances between the equipment, the complexity of the piping systems is high concerning object recognition. Steel structure and piping supports are integrated with the piping system. They are very close.

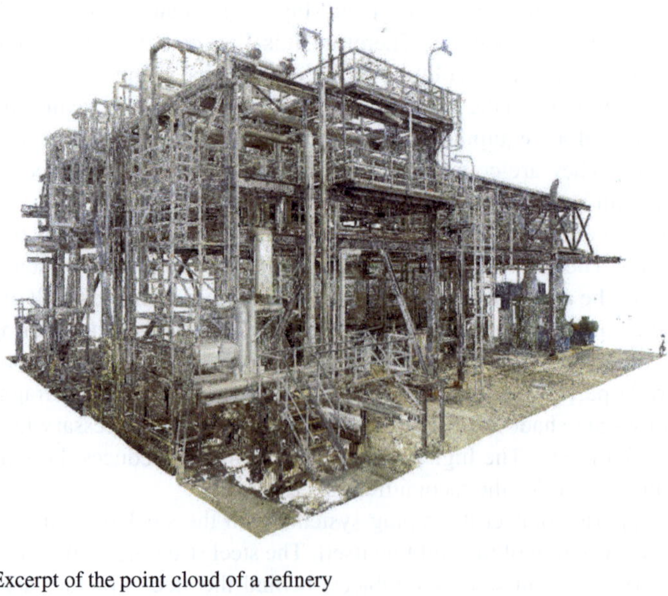

Fig. 8.5 Excerpt of the point cloud of a refinery

Due to the close packaging of pipes and steel structure workers are almost not able to reach each point easily in the refinery through the platforms. These platforms or levels are made of steel grids. They can be reached by ladders or stairs. Railings protect the workers from stepping over the edges of the platforms and protect them from falling.

The sample of a refinery is provided in Fig. 8.5. Pipe strings that can reach several hundred meters are typical for refineries. They are led and supported by pipe bridges, which are basically steel structures in which the piping system takes its course through the space. This compact arrangement of the piping system causes a high density of pipes, which is hurdling the scanning process. Scan shadows and fragmented point clouds respectively reduce the scan quality. Comprehensive object recognition cannot be guaranteed.

Also here the piping system is organized throughout various levels. These can be reached by stairs or ladders, which are attached to the steel structure of the piping bridges. All the mentioned components lead to shadows, which again affect the coloring. Further, the high density of piping and the construction elements lead to high fragmentation and reduced scan quality. Nevertheless, in the presented example, a good scan quality has been achieved.

At the right part of the scan, three pumps can be found. They built redundant systems to ensure the continuity of the processes. However, the identical pumps need to be distinguished and connected to the correct pipe string.

In the background (not visible) huge tanks can be found. Large equipment for storing material or media are typical components in refineries. They often have plenty of incomes and drainages, with various diameters. These have nozzles at their ends and are connected to the piping system. The various diameters need to be determined sufficiently.

Due to the environment, the piping system is occupied with dust and dirt. During the scanning, this is captured as the surfaces. Object recognition must ensure the identification of the correct components even though the exact shape of e.g., a pipe is not detectable anymore.

The long pipes with a heavy dead weight also may lead to sagging of the pipes. They are not ideally straight anymore. However, the object needs to be recognized correctly.

8.3.5 Ship

Even though a ship is not a plant in a closer sense, big ships like cruise liners have complex piping systems as well. These kinds of ships can be seen as small towns on the sea. Starting with the water supply for each area or cabin with fresh water, delivering back grey and black water to internal water treatment systems, or piping system within the machine room, huge complex systems are installed [27].

Figure 8.6 presents the point cloud of a machine room on a ship. Three levels have been captured. On the lower level, the equipment is located. However, the huge electrical engines range from the lower to the middle level. Both levels are separated by floors (green) made of steel. Also, the remaining levels, walls (white), as well as the structure of the ship, are made of steel. Several different piping systems can be found. Some are located close to the roof of each level, others range throughout the different levels. The white-colored pipes are mostly located close to the walls.

Fig. 8.6 Excerpt of the point cloud of a ship (*Source:* HB Hunte Engineering GmbH)

In the middle of the scan can be found free space. To protect crew members from falling such areas are secured by railings. Lamps can be found on each level with sufficient illumination. An exception is the lowest level, which is much darker than the other levels.

From an object recognition perspective, the condition within a ship is challenging. As shown in Fig. 8.6 the pipes range throughout the different levels. Since the floors of each level are made of steel plates, there are breaks or gaps in the pipe strings. Further, the pipes are mostly located close to the walls. This way a 360° capture of the parts is impossible. Many scans need to be carried out and registered together afterward to achieve point cloud useable for object recognition. The challenging scan conditions can also be found in the lowest level. Especially on the left side of the point cloud, many fragmentations can be found. They are caused by scan shadows due to the high density of components.

A ship is a closed space with no natural light inside the machine rooms. The illumination is made by artificial light. Not all levels are illuminated equally. Within the lowest level, almost no illumination can be found. This affects the coloring. Pipes that are painted in the same color seem to change their color throughout their course from the lowest level to the upper level. Because of the changing light conditions, the color varies in the scan.

Pipes with various diameters can be found within the sample. While the smaller diameters are running horizontally the large diameters can be found with pipes running vertically. All pipes are fixed with pipe supports to walls or floors. These additional steel parts also lead to scanning shadows and reduced scan quality. The object recognition process is hurdled this way.

The electrical engines in the lower part of the point cloud belong also to redundant systems comparable to redundant systems of process plants. Here as well the correct linkage of the according pipe systems is crucial.

8.3.6 Discussion—Impact of Complexity Dimensions

Within Sects. 8.3.1–8.3.5, different examples have been presented according to their complexity concerning object recognition. Several factors affect sufficient object recognition. In addition, domain-specific aspects come into place. All have been presented in detail in the mentioned sections and are concluded in the following.

Table 8.2 presents an overview of the dimensions of complexity concerning the discussed examples of real process plants [28]. Four examples are taken from the domain of the process industry: biogas plant, chemical plant, energy plant, and refinery. Additionally, a further however comparable domain has been presented, the shipbuilding industry.

Biogas Plant—Major parts of the piping systems are placed in the open field and only a few are arranged within an overseas container. Therefore, natural light is predominantly. Further, environmental influences by dirt, dust, vapor, or fog can be expected in a point cloud. Due to the size, the complexity is considered as low and the comprehensiveness as high. Also, only a few supporting structures can be found.

Chemical Plant—Since the piping system is placed within a building, only artificial light is present. Further, a high number of supporting structures can be found. They are made of the grid and lead together with the light to many shadows on the piping system. According to the environmental conditions, dirt, dust, and vapor can

Table 8.2 Dimensions of the complexity of analyzed domains

Domain	Light conditions	Environmental influences	Complexity of piping system	Comprehensiveness of scans	Supporting structures
Biogas plant	Natural, partially artificial	Dirt, dust, fog, vapor	Low	High	Low
Chemical plant	Artificial, many shadows	Dirt, durst, vapor	Medium	Medium–high	Medium–high
Power plant	Artificial	Vapor	Medium	Medium–low	High
Refinery	Natural	Dirt, dust, fog, vapor	High	Medium–high	High
Ship	Artificial	Dirt, vapor	Medium	Medium	High

be expected. The complexity of the system is considered medium, while the comprehensiveness of the scan is medium–high, because of less fragmentation within the point cloud. However, many supporting structures such as pipe supports, platforms, or steel construction are present.

Power Plant—The presented example has been scanned within a building. Therefore, artificial light conditions are there. Due to the environment vapor can be expected in relevant amounts. Even though many different pipes can be found the complexity of the piping system is considered as medium. Many scan shadows in terms of fragmentation within the point cloud are recognized, which leads to the comprehensiveness of medium–low. They are caused by too less scan points or the high number of supporting structures and vice versa.

Refinery—Refineries are usually placed in the open field, where natural light is found. In the presented example it is arranged in a predominantly sandy area. Therefore, dirt, dust, fog, and vapor are expectable. Due to size and many auxiliary systems the complexity of a refinery is high. Because of the high density of the pipes, the comprehensiveness is medium. The high number of pipes demands many supporting structures throughout the course of the piping system.

Ship—Piping systems are installed inside ships without any access to natural light. Only dirt can be expected or maybe some vapor. While within the presented the complexity of the system is considered as medium, the entire piping systems within a ship are considered as high. Since many pipes are placed close to the ceiling or walls, the piping system cannot be scanned in a 360° manner. That leads to a comprehensiveness of medium. The supporting structures are considered high since the whole whip is made of steel and the pipes are supported by it.

Even though similarities can be found across the presented domains and examples each comes with its own specific conditions [29]. This ranges from the various light conditions, which can lead also to shadows or the level of contradicting factors due to the environment. Dirt or dust heavily correlates with the area where a piping system is installed. Further, the structural conditions in terms of steel or pipe support have a direct effect on the achievable comprehensiveness of the scan or point cloud. All affect the scan quality and the sufficiency of the recognition process.

The five dimensions may help in the orientation of how challenging the digital twin generation could finally become. Good light conditions, low environmental influences, low complexity, and structures with a high comprehensiveness of the point cloud indicate a tendency for a less challenging project. Conversely, the high complexity and low comprehensiveness of the point cloud are always a particular challenge.

8.4 Generation of Piping CAD Model

As explained in the sections above, the object recognition process highly depends on the scan quality. Environmental conditions, surroundings such as walls or steel structures, as well as light conditions, can significantly influence the quality of a scan. Dirt or dust can change the shape of a pipe component, while the dead weight of a pipe can lead to sagging and a change of appearance respectively. All these facts need to be considered in various ways during the generation process of CAD models.

The 3D scanner captures the surfaces of plants and the piping systems respectively. Due to environmental influences, the shape can vary as explained. Therefore, inconsistencies and deviations can appear concerning the orientation in space. These need to be corrected during post-processing. This way deviations regarding the angles of components need to be corrected. Additionally, misalignments of components might appear for some reasons. These also need to be corrected. This is the basis for the generation of CAD models in CAD systems [11].

After the object recognition within the point cloud, the geometrical and spatial parameters for each component are determined. These parameters are the basis for the generation of CAD models of the piping components [11]. Due to the mentioned influences from the environment or scan quality, inconsistencies appear, as mentioned above. These inconsistencies are corrected in a post-process. Misalignments of components or misinterpreted angles are corrected this way [30]. This way the model undergoes an optimization process that increases the quality of the overall model [31]. Figure 8.7 shows the result. The piping system has been successfully recognized from the point cloud and generated in a CAD system. In order to approve the correctness of the model, they are overlapped with the point cloud in the CAD system.

Fig. 8.7 Overlapping of point cloud and generated CAD model

8.5 Conclusions and Outlook

Object recognition from piping systems in point clouds is a challenging task. The result depends on many various factors. All process plants or piping systems respectively are in an environment [13]. Depending on the domain environmental conditions, vapors, dust, dirt, and more are harming a perfect scan result. Also, the light conditions in terms of brightness, natural light, or artificial light affect the quality of the point cloud. Colors can vary even though it is the same paint or major reflections of highly polished surfaces are contradicting. This directly correlates with domain-specific properties. While some domains such as pharma dedicated to high standards of sanitation refineries are mostly located in a natural environment with dust and dirt [32].

Another factor is the component-specific properties of components. Depending on the application some parts are simple connections between two further components, such as parts, others are made for the change of direction, such as elbows or bends. Specific parameters such as diameters or angles need to be determined to define the recognized components. Due to changed shapes of components e.g., because of dust or dirt the dead weight can lead to sagging, which changes the appearance of components as well [33].

Surroundings such as pipe supports, steel structures, or levels within a plant contradict the scan quality. Often railings of a platform can be found, which is comparable in its appearance to pipes. Or platforms made of steel grid affect shadows on the levels below. Further, they have an influence on the density with regard to packaging. The closer parts are packed together the less comprehensive are the scans. A 360° scan of components is almost impossible in areas with high packaging. The better the scan quality regarding comprehensiveness, the better the result of the object recognition process [34].

When the objects are recognized sufficiently still inconsistencies remain. Misalignment of components or incorrect angles need to be corrected. A post-process for the optimization of results is carried out to achieve better results. Based on that the CAD models can be generated automatically within CAD systems. By overlapping the resulting CAD models with the point cloud, the correctness of the results can be approved [14].

The presented results are sufficient. However, they can be improved. Besides an improvement of the algorithm, the scanning technique appears as a major factor. Today scans are carried out with manual remodeling in mind. Engineers are able to recognize objects correctly with reduced scan quality based on their experiences. An algorithm is limited in this sense. Here comprehensiveness is a major factor. The objects should be as comprehensive as possible (> 80% on average) [10]. By improving the algorithms, it is expected, that the recognition process is getting more robust and can also deal with scans of lower quality. However, by scanning objects for the purpose of automated processing, the results can be improved [35]. Additional effort that might become necessary during the scanning process will be compensated by the automated process later.

Based on the achieved results, the object recognition process is expected to be extended to further areas such as steel structures and intralogistics equipment [36]. Further, additional projects will improve the object recognition process and the results respectively.

References

1. Sutton I (2015) Plant design and operations. Elsevier, Waltham
2. Wossog G (2008) Rohrleitungsbau, 3rd edn. Vulkan-Verlag, Essen
3. Parisher RA, Rhea RA (2021) Pipe drafting and design, 4th edn. Elsevier, Amsterdam. https://doi.org/10.1016/C2019-0-01022-9
4. Selcuk Agca H, Cotone G (2019) Introduction to process plant projects. CRC Press, Boca Raton. https://doi.org/10.1201/9780429466762
5. Uhlenbrock L, Jensch C, Tegtmeier M, Strube J (2020) Digital twin for extraction process design and operation. Processes 8(7):866. https://doi.org/10.3390/pr8070866
6. Thomas CE (2015) Process technology equipment and systems, 4th edn. Cengage Learning, Stamford
7. NN (2022) AVEVA™ asset information management discovery. https://www.aveva.com/en/solutions/operations/asset-visualization/. Accessed 22 Nov 2022
8. Helmus FP (2003) Anlagenplanung: Von der Anfrage bis zur Abnahme. WILEY-VCH Verlag GmbH & Co, KGaA, Weinheim
9. Dal Pont JP, Azzaro-Pantel C (2014) New approaches to the process industries: the manufacturing plant of the future. Wiley, Hoboken
10. Sommer M, Stjepandić J, Stobrawa S, von Soden M (2023) Automated generation of digital twin for a built environment using scan and object detection as input for production planning. J Ind Inf Integr 33:100462. https://doi.org/10.1016/j.jii.2023.100462
11. Grau M, Korol W, Lützenberger J, Stjepandić J (2021) Automated generation of a digital twin of a process plant by using 3D scan and artificial intelligence. Advances in Transdisciplinary Engineering 16:93–102. https://doi.org/10.3233/ATDE210087
12. Sommer M, Seiffert K (2021) Scan methods and tools for reconstruction of built environments as basis for digital twins. In: Stjepandić J et al (eds) DigiTwin: an approach for production process optimization in a built environment. Springer Nature, pp 51–78. https://doi.org/10.1007/978-3-030-77539-1_4
13. Kremer P, Lützenberger J, Müller F, Stjepandić J (2022) An approach for the incremental update of a digital twin of a process plant. Advances in Transdisciplinary Engineering 28:310–319. https://doi.org/10.3233/ATDE220660
14. Sierla S, Azangoo M, Rainio K, Papakonstantinou N, Fay A, Honkamaa P, Vyatkin V (2022) Roadmap to semi-automatic generation of digital twins for brownfield process plants. J Ind Inf Integr 27:100282. https://doi.org/10.1016/j.jii.2021.100282
15. Bondar S, Korol W, Stjepandić J (2022) Object recognition findings in a built environment. In: Stjepandić J et al (eds) DigiTwin: an approach for production process optimization in a built environment. Springer Nature, pp 155–179. https://doi.org/10.1007/978-3-030-77539-1_8
16. Bindel T, Hofmann D (2016) Aufbau des R&I-Fließschemas nach DIN 19227 bzw. DIN EN 62424. Springer, Berlin. https://doi.org/10.1007/978-3-658-15559-9
17. Basu S, Debnath AK (2019) Power plant instrumentation and control handbook: a guide to thermal power plants. Elsevier, London
18. Houson I (2011) Process understanding: for scale-up and manufacture of active ingredients. Wiley, Hoboken
19. Ben-Daya M, Kumar U, Murthy DNP (2016) Introduction to maintenance engineering: modeling, optimization, and management. Wiley, Hoboken. https://doi.org/10.1002/9781118926581
20. Hohmann L, Kössl K, Kockmann N, Schembecker G, Bramsiepe C (2017) Modules in process industry—a life cycle definition. Chem Eng Process 111:115–126. https://doi.org/10.1016/j.cep.2016.09.017
21. Akbulut A, Arslan O, Arat H, Erbaş O (2021) Important aspects for the planning of biogas energy plants: Malatya case study. Case Studies in Thermal Engineering 26:101076. https://doi.org/10.1016/j.csite.2021.101076

22. Moon D, Chung S, Kwon S, Seo J, Shin J (2019) Comparison and utilization of point cloud generated from photogrammetry and laser scanning: 3D world model for smart heavy equipment planning. Autom Constr 98:322–331. https://doi.org/10.1016/j.autcon.2018.07.020
23. Anbari E, Parvini M (2022) Process plant layout optimization considering domino effects and safety devices in a large-scale chemical plant. Comput Chem Eng 167:108006. https://doi.org/10.1016/j.compchemeng.2022.108006
24. Martínez GS, Sierla SA, Karhela TA, Lappalainen J, Vyatkin V (2018) Automatic generation of a high-fidelity dynamic thermal-hydraulic process simulation model from a 3D plant model. IEEE Access 6:45217–45232. https://doi.org/10.1109/ACCESS.2018.2865206
25. Shcherbatov I, Agibalov V, Dolgsuhev A, Belov M (2022) Subsystem for building a digital twin of the main and auxiliary equipment of thermal scheme of thermal power plant. In: Kravets AG, Bolshakov AA, Shcherbakov M (eds) Society 5.0: human-centered society challenges and solutions. Studies in systems, decision and control, vol 416. Springer, Cham. https://doi.org/10.1007/978-3-030-95112-2_20
26. Maalek R, Lichti DD, Walker R, Bhavnani A, Ruwanpura JY (2019) Extraction of pipes and flanges from point clouds for automated verification of pre-fabricated modules in oil and gas refinery projects. Autom Constr 103:150–167. https://doi.org/10.1016/j.autcon.2019.03.013
27. Assani N, Matić P, Katalinić M (2022) Ship's digital twin—a review of modelling challenges and applications. Appl Sci 12:6039. https://doi.org/10.3390/app12126039
28. Hollnagel E (2012) Coping with complexity: past, present and future. Cogn Tech Work 14:199–205. https://doi.org/10.1007/s10111-011-0202-7
29. King PL (2019) Lean for the process industries: dealing with complexity, 2nd edn. Taylor & Francis, New York. https://doi.org/10.4324/9780429400155
30. Latifi SE, Mohammadi E, Khakzad N (2017) Process plant layout optimization with uncertainty and considering risk. Comput Chem Eng 106:224–242. https://doi.org/10.1016/j.compchemeng.2017.05.022
31. Besbes M, Zolghadri M, Costa Affonso R, Masmoudi F, Haddar M (2020) 3D facility layout problem. J Intell Manuf. https://doi.org/10.1007/s10845-020-01603-z
32. Bamberg A, Urbas L, Bröcker S, Bortz M, Kockmann N (2021) The digital twin—your ingenious companion for process engineering and smart production. Chemical Engineering Technology 44:954–961. https://doi.org/10.1002/ceat.202000562
33. Davila Delgado JM, Oyedele L (2021) Digital twins for the built environment: learning from conceptual and process models in manufacturing. Advanced Engineering Informatics 49:101332: https://doi.org/10.1016/j.aei.2021.101332
34. Confalone GC, Smits J, Kinnare T (2023) 3D scanning for advanced manufacturing, design, and construction. Wiley, Hoboken. https://doi.org/10.1002/9781119758532
35. Stjepandić J, Sommer M, Denkena B (2022) DigiTwin: an approach for production process optimization in a built environment. Springer International Publishing Switzerland. https://doi.org/10.1007/978-3-030-77539-1
36. Li D, Liu J, Zeng Y, Cheng G, Dong B, Chen YF (2022) 3D model-based scan planning for space frame structures considering site conditions. Autom Constr 140:104363. https://doi.org/10.1016/j.autcon.2022.104363

Chapter 9
Creation of a New Offering: Digital Twin as a Service

Abstract Plants in the process industry have long-lasting lifecycles of several decades. These long operation phases can only be achieved when these assets undergo continuous maintenance and repair activities. Further, the modernization of plants becomes necessary over time. Therefore the as-is state of the specific brownfield plant needs to be digitized. While the capturing can be automated by laser scanning technology, remodeling is still a manual task. In addition, virtual representation and keeping it up-to-date becomes a demand in the process industry. Based on customer demands, new offerings for the generation of a digital twin of a process plant have been shaped. Input for the offerings are point clouds as a result of laser scans of the specific asset. Depending on the size or complexity of these, the data volume increases drastically and decreases the manageability within CAD or authoring systems respectively. The offer of reduced point clouds addresses this use case. By reducing the number of points the data volume is reduced as well, which leads to better performance within the mentioned systems. In addition, nowadays CAD systems do not provide filtering according to point clouds. The basic segmentation comprises recognized objects in the categories pipeline, equipment as well as structure and provides these as single point clouds. All together represent the entire scan, while the various categories/point clouds can be made visible or invisible. An intelligent filtering can be applied this way. This can also be achieved on component level. The recognized objects are categorized as pipe, elbow, tee, reducer, valve, or flange and are provided as single point clouds as well (Segmentation Plus). By clustering the points that describe one pipeline component, the structure can be broken down on component level. Geometrical as well as spatial information can be derived. Based on that information, CAD models can be generated (CAD Basic). These models can be provided in the native format of specific CAD systems or neutral formats like step or IFC with regards to BIM.

Keywords Digital Twin · Product Lifecycle Management · New offerings · Object recognition from point cloud · Process industry · Auto modeling · CAD generation

J. Stjepandić et al., *Generation and Update of a Digital Twin in a Process Plant*,
https://doi.org/10.1007/978-3-031-47316-6_9

9.1 Introduction

Without the right research and preparation, entering new markets can be challenging. An appropriate entry strategy must be determined in order to increase the company's position in a local or global market or to understand how to introduce a new product into an existing market. The profitability and long-term viability of the company depend on this plan [1].

Product, service, and pricing make up a market offering's three primary elements. Companies in a market may provide a variety of goods and services, but they also add value to the market as a whole. A value proposition is a succinct description of the advantages that customers who purchase a company's goods or services would experience [2].

The lifecycle of a service is similar to that of a product and can be structured correspondingly. Services can be examined by concentrating on how business operations are organized and referring to them as processes. Similar to experiences that happen over time, services are defined as things that need to be organized through a series of interactions between service providers and clients. The coordination of the backstage of services, or the design of the facilities, servers, equipment, and other resources required to generate services, has also been included in the definition of service design [3].

A subscription is the commercial realization of a service. It comprises a contract that grants an enterprise the "right to use" resources from a certain provider. The contract is closed for either a predetermined amount of delivery or a period with a price agreed in advance. The customer mostly has the option to either return the assets at the conclusion of the contractual period or continue paying for and using them.

Service contracts are often managed by third parties. Such service partners are also in charge of getting payments from the end customers. This partner also assumes the risk associated with the customer's solvency. Service agreements are frequently tailor-made. In order to exactly fit the needs of the consumers, they offer fine customized services (maintenance, installation, delivery, upgrade, disposal, customer support, etc.). Services may be offered in cases with complicated handling where specific skills are required for the staff [4]. In the case of business-to-business (B2B) services, all activities are inherently accommodated with the specific industrial environment (e.g., process industry).

The lifecycles of process plants in the process industry may last several decades. To achieve such long operation phases they are undergoing maintenance, repair, and modernization on an ongoing basis. In this sense, maintenance, and repair are applied to ensure process continuity and avoid unexpected downtimes [5]. However, operation of several decades requires major changes in the plants e.g. to increase efficiency or to adapt the processes. Therefore, parts of a plant are kept while others are newly planned [6].

To ensure that newly planned and kept parts fit seamlessly together, the planning of the new parts needs to be done based on the as-is state of the brownfield plant [7].

A digital three-dimensional representation of the actual state is needed [8]. State-of-the-art for capturing the as-is state is the application of a 3D laser scan. Every time the laser hits a surface a point is stored. The result is a cloud of points representing the surfaces of a plant in its current status. By producing several scans from various positions and merging them, a plant can be digitized in a comprehensive manner [9]. This is the basis for the manual remodeling process of a digital three-dimensional representation [10].

For the generation of the 3D digital twin of a plant, the engineer loads the point cloud into the preferred CAD system. This comes with two major challenges. Firstly, with an increasing number of scans also the amount of points is increasing. This leads to an enormous data volume that can reach up to several terabytes according to the size of a plant. Point clouds with these enormous amounts of data are only hardly manageable by nowadays CAD systems. Secondly, nowadays systems do not provide intelligent filtering mechanisms to make parts of the scan visible or invisible according to functional aspects for example [11]. To overcome this hurdle, engineers cut out the areas they are focusing on. This way a reduction of data volume is achieved and better manageability respectively. Further, the complexity is reduced, at least temporarily to the area of interest. This enables the engineer to start the remodeling process [12].

In the first step, the engineers place the large-scale components, the so-called equipment. Components such as reactors, mixers, pumps, heat exchangers, tanks, boilers, etc. are dedicated to this group of components. The nozzles of these components are the connection points to the piping system. In the second step, the armatures and instruments are placed. Finally, the piping components are placed. No matter which class a component is dedicated to, all components are placed within the point cloud in a way, that the points and the surfaces of the digital models are overlapping [13]. The described process is presented in a simplified and generalized manner in Fig. 9.1.

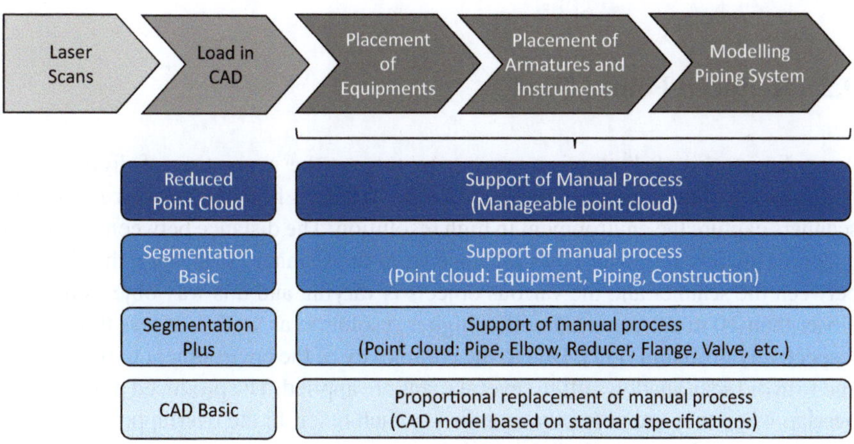

Fig. 9.1 Generalized process of manual and automated remodeling

This manual remodeling of the piping system is necessary to ensure, that newly planned and kept parts are fitting seamlessly together, as stated above. In this sense, the 3D digital twin is the basis for the planning process of the new parts [14]. E.g. the parts to be replaced are cut out of the as-is model and the new part can be planned and designed according to the processual pre-conditions and the geometrical and spatial conditions of the plant [15].

Remodeling the as-is state of plants is a time-consuming task. Operators have to spend significant investments for digital twins of their assets. Pre-condition is a reasonable return on investment. This usually limits the use of digital twins to modernization tasks, which enable e.g. a reasonable increase in efficiency. Even though maintenance and repair can benefit of three-dimensional digital twins, the monetary value often is not high enough to generate digital twins only for the purpose of maintenance or repair [16].

At the top of Fig. 9.1, the already described manual remodeling process is presented in a simplified and generalized manner. For various reasons, the fully manual generation of 3D digital twins is too expensive and takes too long time. A cost-efficient solution is needed that supports the modeling process significantly to make the use of digital twins economical, also for certain tasks beyond modernization. Adequate support can be achieved in various ways. The alternatives are presented in the lower part of Fig. 9.1 and condense the new offerings from a reduced point cloud up to a simplified CAD model in four consecutive steps that build on each other: Reduced Point Cloud, Segmentation Basic, Segmentation Plus, and CAD Basic. All offers are introduced in detail within the following sections.

In this chapter, essential components of a new offering to generate the digital twin with customers' benefits are presented. Section 9.2 explores the characteristics and benefits of a reduced point cloud, followed by Sect. 9.3 on Segmentation Basic. Section 9.4 explains the advanced segmentation defined as Segmentation Plus. The final offering of CAD Basic is discussed in Sect. 9.5. The chapter ends with a conclusion and outlook.

9.2 Reduced Point Cloud

According to Fig. 9.1, the first manner to support the generation of digital twins is a reduced data volume of the point cloud. Modern high-performance 3D laser scanners capture the environment in high resolution. The distance between the points is approximately 6 mm in 10 m distance from the scanner [17]. Since the distance between the scanner and the various objects is varying and this way objects that are closer than 10 m are captured with a higher resolution as well. It is less than 6 mm respectively (1–3 mm) [17]. Due to the complexity of the environment to be scanned and to avoid scan shadows often several scans are applied. The produced point clouds overlap when they are correctly arranged to each other. In the overlapping are more points of several scans included, which increases the resolution as well.

Fig. 9.2 Energy plant—Reduced Point Cloud

With such high resolutions point clouds look like super realistic 3D pictures. Even though a higher resolution may be beneficial for the engineers during the manual remodeling process in terms of recognition of components, it leads to a higher data volume to a certain extent. According to the size of the scanned area and the number of points respectively merged scans can lead to data sizes of terabytes. This huge data is hardly manageable in contemporary CAD systems.

To cope with the challenge of the high data volume, the first offering is a reduction of volume by filtering out unnecessary points. The number of points is reduced to such an extent, that ensures object recognition by the engineers with reasonable effort. At the same time, the reduction of points leads to a reduced data volume without losing the information worthiness. By increasing the point distance to approx. 20mm a reduction in terms of the data size of approx. > 90% can be achieved. Figure 9.2 shows parts of the scan of a public energy plant. The volume is 20 × 17 × 8 m (length × width × height) and the original point cloud consists of 76.46 million points. By reducing the number of points by approx. 94% of the data size has been reduced from 1.72 GB to approx. 100 MB.

9.3 Segmentation Basic

In order to generate a 3D digital twin out of a point cloud the engineers load the point cloud into the specific CAD system in use. It provides the as-is state of a plant in terms of the external shape [11] and covers equipment, piping, building, structure, floor, etc. The engineers who are remodeling the plant aim to place the components into the point cloud in a way, that the placed component models overlap with the point cloud. Both surfaces should be mapped most ideally.

Besides the piping components also other surroundings such as pipe supports or steel structures etc. are captured by the laser scanner as well. Pipe bridges are prominent examples of areas with many pipe components closely placed. In addition, the pipes are fixed by piping components, which are connected to the steel structure. Achieving an overview and placing the components correctly is difficult for the engineers remodeling the plants [15].

A mechanism enabling the engineers to filter the point cloud according to the technical functionality of objects would provide significant benefits here. However, nowadays CAD systems lack in terms of such filtering. To cope with such situations and challenges engineers are forced to cut out parts of the point cloud. This way they can extract parts like building structures to achieve a better view of specific areas. Conversely, intelligent filtering by segmentation of object categories in a plant enables the engineers to make parts of the point cloud visible or invisible, which is needed to increase the efficiency of manual remodeling.

Figure 9.3 shows the result of the first segmentation step. The artificial intelligence (AI) algorithms analyzed the point cloud in order to recognize objects. To cope with the complexity of the plants or the point clouds respectively, the first semantic segmentation process distinguishes between equipment (yellow), pipeline (green), and structure (blue). The structure covers steel structures, building structures, and other objects, which do not correlate with the categories of equipment or piping. There can be any number of categories [18].

Fig. 9.3 Energy plant—Segmentation Basic (equipment, piping and construction)

All points are assigned to one of these categories and each category is stored in a single point cloud. The engineers can load the three point clouds into their CAD system and can selectively make them visible or invisible. This way filtering can be applied which enables the engineer to focus, for instance, e.g. on the piping system without having the structure visible at the same time. By making the other parts of the plant visible again, the context may be given again. The offering comprises three point clouds according to the mentioned categories in reduced form in terms of data volume. Furthermore, they can be provided in various formats such as *.ply, *.e57, etc.

9.4 Segmentation Plus

As mentioned before in Sect. 9.3, a differentiation between classes or groups of objects provides benefits in terms of a filtering mechanism applied by making the various point clouds visible or invisible. Being able to filter the point model according to functionality increases the efficiency of the remodeling process. However, when the packaging of piping components is very close even filtering between structure, piping, and equipment might not be sufficient. In areas in which many piping components (reducer, pipe, elbow, tee, etc.) are close to each other filtering on a more detailed level is needed.

The point cloud covering the piping of the first segmentation is the basis for the second segmentation process [19]. Again, AI-based algorithms run an object recognition process to distinguish between different component types:

- Elbow
- Flange
- Pipe
- Reducer
- Tee
- Valve

Figure 9.4 shows the result of the second segmentation process. The mentioned components are highlighted in various colors. Each component type can be provided in a single point cloud. Analog to the offering "Segmentation Basic" a filtering mechanism can be applied by loading all point clouds into a CAD system and making specific point clouds visible or invisible. Detailed filtering on the component level can be achieved. The offering of "Segmentation Plus" includes the point clouds of Segmentation Basic and the point clouds of Segmentation Plus. By replacing the pipeline point cloud of Segmentation Basic with the point cloud bundle of Segmentation Plus, the engineers can achieve filtering and structuring of the whole point cloud to the component level. All point clouds are available with reduced data volume.

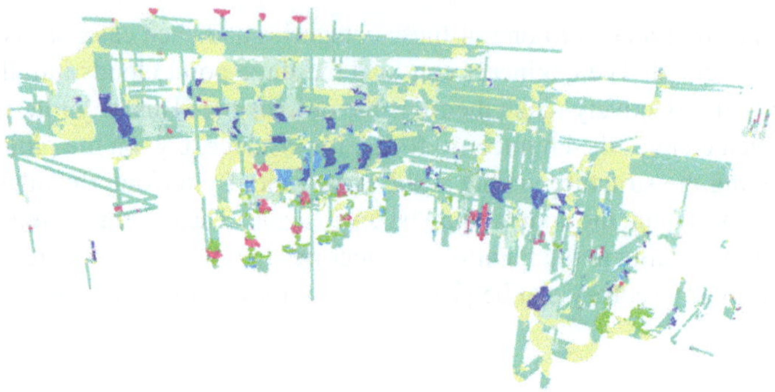

Fig. 9.4 Energy plant—Segmentation Plus (pipe, elbow, reducer, tee, valve, etc.)

9.5 CAD Basic

When it comes to large-scale modernization projects with significant changes, major parts of a plant or even the entire plant need to be remodeled. Depending on the size, this can take several months of work with several engineers. Due to the high effort and costs respectively, the generation is often only done for projects with a reasonable return on investment, like modernization projects.

Such projects need a basis for planning the modernization. In this sense, the generated 3D digital twin is necessary. However, only the basis for carrying out the most important work: planning of the to-be-modernized units and parts. Often the same engineers responsible for the planning also do the remodeling. This way they spent capabilities and resources to a certain degree on tasks not correlating to their level of education. These well-paid experts are blocked from tasks that require their expertise and come with higher hourly pay.

An automated generation of a 3D digital twin will provide benefits in several manners. Firstly, the elapsed time for the generation can be reduced significantly. Secondly, the cost will be reduced respectively to a reasonable degree. Finally, the engineers can focus on their core competencies according to their education, the planning of new plants, or at least new parts of the plants.

Depending on the application case the requirements for accuracy can vary. For major modernization tasks, it needs to be ensured that the kept and the new parts fit seamlessly together. Therefore, the CAD model has to be built on the specification of the piping classes. They define the specific attributes of available components to be applied [15].

Another example is the exchange of huge components. In many cases, they have been initially installed and the surrounding piping system evolved. New parts and new pipes have been installed into the available space. In case these huge components must be replaced, it needs to be ensured beforehand, that the old component will fit through the piping system to be extracted. In such cases, a simple 3D model is needed to reflect the external shape of the as-is piping system. A dedicated specification is not applied in such cases.

Figure 9.5 shows an IFC model [20] generated in an automated manner. To achieve this goal after the second segmentation step, clustering, which clusters the number of points representing the surface of one component, is applied. Based on the result geometrical and spatial parameters can be derived. At this stage, optimization is necessary to improve the gained results:

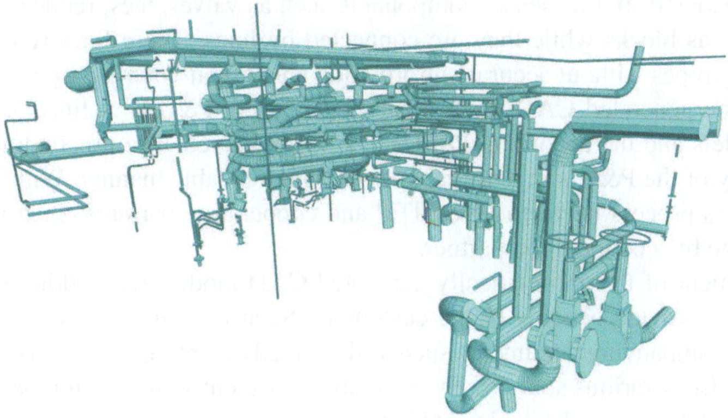

Fig. 9.5 Energy plant—CAD Basic

1. Misalignment of piping components
2. Variation of the derived diameters of different components
3. Angle differences between components

The optimized results are mapped to components of the piping classes applied at the specific plant. According to the derived spatial information, the components can be placed in space. Generated CAD models can be provided in native formats of the relevant CAD systems for plant engineering or neutral formats such as *.step, *.ifc, *.3dxf, *.JT and further. Both the models as well as the point clouds can be loaded into the CAD systems. This enables engineers to double-check the models or add components not recognized by the AI (e.g. due to scan shadows).

9.6 Optional Service

Depending on the use case high accuracy of the model is of relevance. CAD Basic provides a CAD model, which represents the as-is state of a plant. However, it may lack in terms of accurate metadata. No matter which kind of scan is applied, it is only possible to capture the surfaces. During the process of automated remodeling, the structure is reconstructed sufficiently, however, meta information such as a precise nominal diameter, the use of insulation, or specific part numbers are not available.

To enrich the CAD model with additional information, the quality of such models can be improved significantly. If high accuracy is needed, information such as piping class, part number, etc. is added manually by the design engineer. That information is derived from the accordingly P&IDs. These representations of a piping system reflect the schematic structure of a plant. It comprises equipment, fittings, and instrumentations of a piping system as well as the logical connection of them. 3D information is not considered. In this sense, components such as valves, tees, reducers, etc. are considered as blocks while they are connected by lines. These lines represent the connecting pipes without accurate information about their course in space.

Also, the generated CAD models can benefit from P&IDs. A linkage between CAD models and the according P&IDs can be automated. The precondition is the availability of the P&IDs in a digital and machine-readable manner. With regard to the business process between PROSTEP and cooperation partners, such a linkage can be done by a cooperation partner.

Enrichment of the automatically generated CAD models with additional information will provide benefits to the customers. Such information may come from directly accompanying documents such as the already mentioned P&ID but can also be derived from various sources. By integrating different sources with the models a holistic digital twin can be achieved [21].

9.7 Conclusions and Outlook

Digital twins are virtual representations of real existing objects/products. In the process industry, the generation of digital twins covering 3D information are generated manually so far. Especially with regard to planning tasks for brownfield plants, these digital models are of relevance. They build the basis for planning modernization for example.

To capture the as-is state and to digitize plants, 3D laser scanning is state-of-the-art technology. Outcome is a point cloud representing the surfaces of the scanned components. The scans are carried out from various positions in the plant to avoid scan shadows and ensure that all important areas are covered.

The received point cloud is the basis for the remodeling of the plant models. This manual carried out remodeling is a time-consuming task that leads to high costs respectively. It usually is carried out by engineers that are well educated and come with a high hourly rate. This drives the costs additionally.

For well-educated engineers, such redesign tasks are not aligned with their level of education. By taking over such remodeling tasks they are hindered from tasks of higher value such as planning tasks which require expert knowledge on the one hand and come with a certain degree of innovational challenges on the other hand.

Further, according to the complexity and the size of a plant, point clouds can consist of billions of points. This leads to point clouds with high data volume. Nowadays, CAD systems are reaching their limits by processing these data amounts. Engineers are treated to cut out of the point cloud areas they want to focus on. The context is lost for the moment, which hurdles the work of the engineers to some extent.

Even though digital twins provide a broad range of benefits to designers and operators, their use is limited to cases with a reasonable return on investment. The enormous costs of generating such a digital twin reduce the number of reasonable cases enormously. Typically modernization activities provide such a return on investment, while maintenance activities often do not. Uncertainties in the planning of the tasks must be compensated by additional personnel or extra effort in terms of time.

The introduced solution enables an automated generation of digital twins in an automated manner. Based on artificial intelligence, object recognition from the point cloud is enabled. Pipe components can be detected and spatial as well as geometrical information can be derived. This enables a wide range of available services: Reduced Point Cloud, Segmentation Basic, Segmentation Plus, and CAD Basic.

By reducing the number of points to a reasonable data volume, the performance issues of nowadays CAD systems can be intercepted. A reduced point cloud facilitates the handling of point clouds of complex plants. Remodeling work is getting more convenient and efficient this way.

The basic segmentation separates the point clouds into equipment, pipeline, and surroundings. Each class can be stored in a single-point cloud. Engineers can load the point clouds into their CAD system and enable filtering by making the different point clouds visible or invisible.

Segmentation Plus distinguishes the point cloud of the pipeline on component level. In addition, all point clouds from segmentation basic are provided. The customers are enabled to apply filtering by making the various point clouds visible or invisible on component level. Straight pipes, elbows, tees, valves, reducers, and flanges can be filtered that way.

Finally, CAD Basic provides CAD models in various native or neutral data formats according to customer needs. Based on the results of the clustering process, geometrical and spatial information is derived and the models are built. The engineers can load the models as well as the dedicated point clouds together in the specific CAD system and can apply a double check here.

For the time, the development focuses on the object recognition of the piping system. However, there is an additional remodeling task that can be automated as well. The basic structure of plants is usually made of steel. In the context of modernization, these steel parts need to be remodeled as well. An automated remodeling of steel parts would provide additional benefits to the customer and will provide additional business cases and offerings in the future.

Even though the piping systems are built of standard components, the number of combinations is infinite. It is expected, that new projects or situations will further appear and additional training of the AI algorithms will be necessary. The improvement of the data models and further training respectively is a key factor for a successful business in the future.

The object recognition of piping components is limited in terms of accuracy. All parameters are derived from the scanned surfaces. Factors such as scan quality, dust, or noise can reduce the outcome. Additional information can be derived from the according P&IDs. For that reason, a linkage to the P&IDs enables an enhancement of the models with meta information stored in the point clouds.

Also, the current business models focus on the domain of the process industry. However, other industries can benefit from the automated remodeling of piping systems. The shipbuilding industry is comparable to some extent to the plant industry in terms of piping systems. Applying the technology e.g. to shipbuilding will open new markets for the offerings already introduced.

The introduced offering provides significant advantages to customers. A linkage with the according P&IDs as envisaged will provide additional benefits and enable new business models in the future [22]. At the same time, the offerings can be brought to further markets and branches which comes with a high economic potential for the future.

References

1. Lymbersky C (2008) Market entry strategies: text, cases and readings in market entry management. Management Laboratory Press, Hamburg
2. Osterwalder A, Pigneur Y, Bernarda G, Smith A, Papadakos T (2014) Value proposition design: how to create products and services customers want. John Wiley & Co, Hoboken
3. Morelli N, de Götzen A, Simeone L (2021) Service design capabilities. Springer Nature Switzerland AG, Cham. https://doi.org/10.1007/978-3-030-56282-3
4. Stjepandić J, Sommer M, Stobrawa S (2022) The commercialization of digital twin by an extension of a business ecosystem. In: Stjepandić J et al (eds) DigiTwin: an approach for production process optimization in a built environment. Springer Nature, pp 205–233. https://doi.org/10.1007/978-3-030-77539-1_10
5. Lock D, Wagner R (eds) (2018) The handbook of project portfolio management. Routledge, London. https://doi.org/10.4324/9781315206592
6. Houson I (2011) Process understanding: for scale-up and manufacture of active ingredients. Wiley, Hoboken
7. Sutton I (2015) Plant design and operations. Elsevier, Waltham
8. Selcuk Agca H, Cotone G (2019) Introduction to process plant projects. CRC Press, Boca Raton
9. Sommer M, Seiffert K (2021) Scan methods and tools for reconstruction of built environments as basis for digital twins. In: Stjepandić J et al (eds) DigiTwin: an approach for production process optimization in a built environment. Springer Nature, pp 51–78. https://doi.org/10.1007/978-3-030-77539-1_4
10. Stjepandic J, Sommer M, Denkena B (2022) DigiTwin: an approach for production process optimization in a built environment. Springer International Publishing Switzerland. https://doi.org/10.1007/978-3-030-77539-1
11. Grau M, Korol W, Lützenberger J, Stjepandić J (2021) Automated generation of a digital twin of a process plant by using 3D scan and artificial intelligence. Advances in Transdisciplinary Engineering 16:93–102. https://doi.org/10.3233/ATDE210087
12. Kremer P, Lützenberger J, Müller F, Stjepandić J (2022) An approach for the incremental update of a digital twin of a process plant. Advances in Transdisciplinary Engineering 28:310–319. https://doi.org/10.3233/ATDE220660

13. Löwen U, El Sakka F, Schertl A, Fay A (2020) A systematic approach how to build, maintain and use an integrated plant model. Automatisierungstechnik 68(6):423–434. https://doi.org/10.1515/auto-2019-0097
14. Sierla S, Azangoo M, Fay A, Vyatkin V, Papakonstantinou N (2020) Integrating 2D and 3D digital plant information towards automatic generation of digital twins. In: 2020 IEEE 29th international symposium on industrial electronics (ISIE). https://doi.org/10.1109/ISIE45063.2020.9152371
15. Parisher RA, Rhea RA (2021) Pipe drafting and design, 4th edn, Elsevier, Amsterdam. https://doi.org/10.1016/C2019-0-01022-9
16. Kockmann N, Bittorf L, Krieger W, Reichmann F, Schmalenberg M, Soboll S (2018) Smart equipment—a perspective paper. Chem Ing Tec 90(11):1806–1822. https://doi.org/10.1002/cite.201800020
17. NN (2018) Leitfaden Laserscanning im Anlagenbau Vorbereitung und Durchführung von Laserscanningprojekten für die Planung und Dokumentation von industriellen Anlagen, 5. Auflage, Fraunhofer-Institut für Fabrikbetrieb und -automatisierung IFF, 2018. https://www.iff.fraunhofer.de/content/dam/iff/de/dokumente/publikationen/leitfaden-laserscanning-im-anlagenbau.pdf. Accessed 10 Jan 2023
18. Stjepandić J, Sommer M (2022) Object recognition methods in a built environment. In: Stjepandić J, Sommer M, Denkena B (eds) DigiTwin: an approach for production process optimization in a built environment. Springer series in advanced manufacturing. Springer, Cham, pp 103–134. https://doi.org/10.1007/978-3-030-77539-1_6
19. Stjepandic J, Bondar S, Korol W (2022) Object recognition findings in a built environment. In: Stjepandić J, Sommer M, Denkena B (eds) DigiTwin: an approach for production process optimization in a built environment. Springer, Cham, pp 155–179. https://doi.org/10.1007/978-3-030-77539-1_8
20. Yacout S, Ebrahimipour V (2015) ISO 15926. In: Ebrahimipour V, Yacout S (eds) Ontology modeling in physical asset integrity management. Springer, Cham. https://doi.org/10.1007/978-3-319-15326-1_1
21. Bamberg A, Urbas L, Bröcker S, Bortz M, Kockmann N (2021) The digital twin—your ingenious companion for process engineering and smart production. Chemical Engineering Technology 44:954–961. https://doi.org/10.1002/ceat.202000562
22. Martínez GS, Sierla SA, Karhela TA, Lappalainen J, Vyatkin V (2018) Automatic generation of a high-fidelity dynamic thermal-hydraulic process simulation model from a 3D plant model. IEEE Access 6:45217–45232. https://doi.org/10.1109/ACCESS.2018.2865206

Chapter 10
Digital Twin: Conclusion and Future Trends in Process Plants

Abstract The generation of the digital twin for a brownfield process plant can be offered in the market as a service provided in collaboration with a few partners. In this book, we have tackled specific technical and economic issues, benefits, and limitations of a dedicated process to achieve this objective. It is a service with inherent self-improvement, while it facilitates a fast feed-forward of new software solutions, initiated by advances in research, to a daily-used industrial service. In this chapter, we draw the way for the further development of the DigiTwin solution which is either obvious or can be expected. Moreover, we highlight some still open challenges with respect to the digital twin and its associated research (sub)domains in a brief overview. As in the preceding chapters, a consistent taxonomy with a self-explainable set of eight dimensions that can be articulated with a limited number of characteristics is reused for the assessment and positioning of current trends and challenges in digital twins. Four dimensions have been identified with potential for short-term advances: instantiation (reusability) of a digital twin, dealing with different accuracy of singular components, and improved processing of raw data. External impacts due to advances in the scanning process, process modeling, as well as data processing and services, are explored too. A section is dedicated to the synchronization with the piping and instrumentation diagram which is identified as a key improvement for digital twin. In the discussion section, ten main challenges identified in the recent publications are addressed. Apart from the functional improvements, the handling of requirements, traceability, and metrics for the assessment of digitization pose the main challenges. Furthermore, the challenges in architecture, data processing, and automated simulation workflow round up this section. Intellectual property and its protection are highlighted in the context of a collaborative environment. The results show an evolution of digital twin's role from an enabler of cyber-physical systems to a constituent of Asset Lifecycle Management, the central platform for data integration and processing. Finally, the conclusion section emphasizes again the importance of seamless interoperability.

Keywords Digital Twin · Taxonomy · Brownfield plant · Piping engineering · P&ID · 3D scan · Object recognition · Artificial Intelligence · Interoperability · CAD · ALM

J. Stjepandić et al., *Generation and Update of a Digital Twin in a Process Plant*,
https://doi.org/10.1007/978-3-031-47316-6_10

10.1 Introduction

A practical approach for the generation of the digital twin and its commercial, productive application in process plants has been presented based on their gradual emergence and discussed at length in the preceding chapters of this book. As a continuation and an extension of the previous research and development [1], we have addressed specific technical and economic issues, benefits and limitations of a dedicated process chain or step in isolation focused on a specific project aim as a part of the corresponding chapters in the discussion sections. Furthermore, this work should facilitate a fast feed-forward of new software solutions to a daily-used industrial service [2].

The purpose of this final chapter is twofold. First, we draw the way for the further development of the solution as presented in this book which may lead to functional and economic improvements, called 3DigitalTwin. Second, we integrate and present current trends and challenges with respect to the digital twin and its associated research (sub)domains, in a particular application of Artificial Intelligence (AI), in a comprehensive overview. To provide this integration in a structured, methodical manner, the impact of the industrial environment, business development, and socio-technical dimensions of digital twin is also explored. With this procedure, authors follow the structure for the final chapter already developed in the prior research on digital twins [1].

A consistent taxonomy with a self-explainable set of dimensions that can be articulated with a limited number of characteristics is helpful for the presentation and discussion of current trends and challenges in digital twins. Furthermore, this provides a basis for the structure, comparison, and assessment of different approaches to the digital twin [3]. With the aid of these dimensions, it is possible to provide generalizations about the results and avoid the typical entanglement of digital twin descriptions with particular sectors of the industry and a wide range of application domains. The eight dimensions used are derived from the taxonomy of digital twins developed by van der Valk [4], previously explored in Sect. 3.2: data link, purpose, conceptual elements, model accuracy, interface, synchronization, data input, and time of creation.

Although not all-inclusive, these eight dimensions together comprise an assessment framework that simplifies the processes involved in configuring, implementing, and exploiting a particular expression of a digital twin in accordance with a certain strategy. This procedure is more concerned with the application than it is with the architecture of a specific solution. The benefits of the particular digital twins can be identified more easily afterward. In order to help distinguish different concepts related to digital twins from one another, a compact taxonomy can make a significant contribution to closing that research gap. Table 10.1 highlights the fulfillment of individual dimensions based on the most recent review which is an extension of Table 3.1 [4].

The characteristics of the digital twin presented in this work are highlighted in light grey in Table 10.1. It works with a one-directional data link, for use in transfer and repository, physically bounded to one physical twin, which works with identical

Table 10.1 Practice of digital twin, derived from taxonomy [4]

Dimension	Characteristics		
Data link	One-directional	Bi-directional	
Purpose	Processing	Transfer	Repository
Conceptual elements	Physically independent	Physically bound	
Accuracy	Identical	Partial	
Interface	M2M	HMI	
Synchronization	With	Without	
Data input	Raw data	Processed data	
Time of creation	Physical part first	Digital part first	Simultaneously

accuracy as the physical twin, controlled by a human–machine interface without synchronization between physical and digital, based on an already existing physical twin.

For this conclusion, we have identified the next possible improvements of the digital twin presented here which are related to three out of eight dimensions (highlighted in dark grey in Table 10.1):

- **Conceptual elements**: Although a digital twin is mostly directly bound to its dedicated physical counterpart in a one-to-one ratio, the occurrence of an independent, updateable, scalable relationship between physical and digital is desirable and meaningful. Such a flexible relationship would make the reuse, incremental update, and upgrade of an existing digital twin possible. In such a case, a digital twin can be seen in combination with other physical systems, or one system can possess multiple digital twins [5–8]. Despite the utilization of standard components and existing templates [9], a standard procedure to draw a digital twin would be very helpful [10].
- **Accuracy**: A physical object is mostly described with an identical accuracy either in a point cloud or in a CAD model, which makes it easy to control the entire process [11]. While accuracy (defined as the distance between two neighbor points) directly determines the data volume of a point cloud, an identical accuracy yields an unnecessary huge volume of data for point clouds. Otherwise, a partial (variable) accuracy makes sense for an efficient representation and fast processing [12–14]. Subsequently, the process of data acquisition and preparation of the point cloud should be transformed to an adjustable, partial accuracy depending on the structure of the object in question and the process requirements.
- **Data input**: Data is the "fuel" of the digital twin [15]. Therefore, digital twins receive input data from sensors or databases, either directly as pure, raw data from data acquisition devices or pre-processed (e.g., by analytic software) before use in the digital twin. Here, a substantial distinction only could be made in a further breakdown of data processing [16–18].

Reuse is a further benefit of a digital twin. A digital twin as described here could be used in an additional instance for further optimization tasks such as layout optimization of a plant. The layout of a process plant is an essential factor in designing a chemical process which can lead to enhancing the performance of the plant as well as reducing economic damages in the case of any incident in the plant. Conversely, an inappropriate plant architecture has a substantial impact on production efficiency, which can lead to a backlog of raw materials in the production process, a wide distance between operations, and low use of production units. Furthermore, the real-time feedback of physical information from the plant can impact possible scenarios including fire, explosion, domino effects, and the existence of safety devices. Nowadays, it is still infrequently taken into consideration by the several optimization methods for plant layouts, making it impossible for the layout to achieve self-optimization while in use [19].

Recent research addressed the plant layout optimization based on digital twin, in which the plant layout problem is solved by twin data fusion, information, and physical interaction fusion, and data analysis and optimization. First, before optimizing the plant partitioning through simulation analysis, a sub-framework for digital twin-based plant partitioning is built. Second, a sub-framework for digital twin-based equipment layout optimization is described, in which the selection of equipment layout is made using real-time data collection and twin data that has undergone value-added processing. Third, a workshop's sub-framework for digital twin-based distribution route optimization is created. The results illustrate that the proposed model can reduce piping costs, land costs, the domino hazard index, and the costs associated with domino escalation by a significant share. In addition, a lack of safety devices can significantly increase the piping cost of the plant [20].

Despite the fact that a digital twin is a highly integrated system (of systems), different integration levels are needed to handle various degrees of technological uncertainty [21]. However, integration issues become more sensitive in implementation projects with a higher level of uncertainty. Digital thread provides a means for continuous interoperability of all digital models over the entire product lifecycle phases. Used straight-forward, this digital thread-driven method models the manufacturing tasks by heterogeneous information network to analyze the product quality information during the manufacturing process and adjusts the subsequent manufacturing tasks according to the analysis results [22]. The collaboration between production units forms a stable and reliable operation mode for improving production efficiency during the whole production process. In addition, there are consequential issues of configuration and risk management [23].

The outline of the chapter is as follows. In Sect. 10.2, the research and development directions of the 3DigitalTwin solution presented in this book are drawn and discussed. In Sect. 10.3 synchronization with piping and instrumentation diagram is drawn, followed by the discussion in Sect. 10.4. Finally, closing remarks and conclusions on the presented work are given in Sect. 10.5.

10.2 Developments with Impact on 3DigitalTwin Solution

This book shows how a digital twin of a process plant can be efficiently created with high-quality, reusable data. The focus of the explained method is on the initial set-up, which means that a digital twin can be created once with the method, but no further adjustment of the digital twin is carried out after the method has been performed.

10.2.1 Advances in the Scanning Process

Despite the fact that the generation of the point cloud of the object in question is not the subject of this work because it is performed by an independent, external capture partner, the way the point cloud is generated plays an important role in the extent of data, its quality and the performance of the entire process. Furthermore, the state of the input data determines the extent and the necessary efforts for interactive, manual preparation of the input data. Therefore, we will briefly discuss some possible directions for future developments in the area of scanning.

Requirements for the recording system have remained stable in recent years. High accuracy can be achieved by a stand-alone terrestrial laser scanner and no significant improvement can be expected. Two further important requirements are fast recording and simple handling. The performance improvements of a terrestrial laser are limited here due to the inherent weakness of its large dimensions, weight, and limited portability [11].

One direction in the development of scanning technology is predefined: Efforts have been made continuously to reduce the size and weight of scanning devices in order to increase their portability. To operate, handheld devices often need to be connected to a laptop. A scanning device needs to combine a laser, computer, storage, and battery in a small, portable package in order to be genuinely independent. The most basic handheld scanning device is a cell phone, which can operate basic scanning software [14]. It can be easily predicted that the standard acquisition device will become a smartphone carried by an adequate carrier (human, tripod, drone), and the functionality will be more and more moved to the corresponding apps.

As discussed in Chaps. 6 and 7, the acquisition of 3D point clouds is frequently hindered by the occlusion of the objects in the scene. To resolve this problem which could make the method described here useless, an approach for optimizing LiDAR (Light Detection and Ranging) surveys using metaheuristics such as local searches and genetic algorithms was developed. The method improves the usability of a LiDAR device by generating a set of optimal scanning locations to densely cover the real-world environment represented through 3D synthetic models. Compared to previous approaches, 3D occlusion is compensated by varying the height of the sensor. In this way, the combination of local searches and genetic algorithms generates a reduced set of locations capable of optimizing the scanning of buildings.

Despite being accelerated in CPU and GPU, the proposed methodology is still time-consuming due to the dimensionality of both 3D environments and LiDAR scans guided by specifications of commercial sensors [24].

10.2.2 Advances in the Process Modeling

As already stated in previous chapters, no standard procedure for the emergence of the digital twin is known at this time. Almost every implementation of the digital twin was realized with its own method and tool set. Therefore, standard components and procedures were used wherever possible, and we conducted requirements and process design by using Cameo software (Sects. 2.2 and 2.3), well-known in systems engineering.

To overcome this inherent hurdle, a system definition based on standards is necessary—preferably in a digital twin-compliant language. The modeling language is used to describe models of different domains or in different forms with a unified representation so that the models can be run in a simulation environment. At least, this language should represent the relevant dimensions of digital twins (Table 10.1) to a sufficient extent. Recent research investigated the modeling languages used to implement digital twins in order to facilitate the integration and cooperation of digital twins that have been independently developed with varying standards, and protocols [10].

In an overview of modeling languages, only Digital Twin Definition Language (DTDL), derived from Azure Digital Twin, suffices for the rudimentary requirements of digital twins [25], in particular for the integration of Internet-of-Things components. Conversely, the high-level modeling language must be able to translate to various low-level modeling languages and be able to accurately describe both the exterior architecture and the interior behaviors of the real system. Additionally, depending on current low-level modeling languages as well as high-level modeling languages, modeling tools can be created [26].

Digital twin modeling is digital modeling in virtual space for the properties, methods, behaviors, and other characteristics of the physical twin. Specifically, digital twin modeling is the core for the accurate portrayal of the physical entity, which enables the digital twin to deliver functional services and satisfy the application requirements. It is provided in six aspects: model construction, model assembly, model fusion, model verification, model modification, and model management. Modeling of the digital twin can be subdivided into six model clusters [27]: hierarchy of the digital twin model, discipline of the digital twin model, dimension of the digital twin model, universality of the digital twin model, functionality of the digital twin model and modeling aspects of the digital twin modeling.

The next step in the generic process modeling is model engineering which provides a complete technical methodology and system to guide activities in the whole lifecycle of a model. Model engineering is defined as the systematic, standardized, quantifiable, and domain-independent engineering methodology that ensures the whole model lifecycle is credible at the lowest cost, which includes theories, methods, technologies, standards, and tools. Model engineering is applied to manage all the data, knowledge, activities, processes, people, and organizations involved in the whole model lifecycle, and considers time, cost, and other metrics of developing and maintaining a model [26].

While a digital twin generally lasts a long lifecycle and keeps continuously evolving with the physical system, an evolutionary concurrent modeling method for a digital twin is proposed based on a framework. Model engineering is aimed at establishing a complete technical methodology and generic system to guide activities in the whole lifecycle of a model. From the point of view of model engineering, the whole lifecycle of a digital twin should be comprehensively studied as one object [26]. This is the point where the digital twin needs to be embedded in overall Product Lifecycle Management (PLM). Moreover, a digital twin applies not only to virtual representation but also to the prediction of expected product operation [28]. The aim of the digital twin modeling method, which is based on PLM theory, is to provide sufficient fundamental data and supporting models for the generation of the product digital twin. PLM uses digital thread technology to build a high-fidelity digital twin for the entire life cycle of the part while modeling and collecting data for the entire process from design to end-of-life [29].

10.2.3 Advances in Data Processing and Services

Large and complex assets (e.g., plants, facilities, infrastructure) are the foundation for production, transport, traffic, and daily life, and they enable the flow of goods, information, and services within industrial areas, urban and regional settings. While they must become more efficient, resilient, and sustainable, data-centric solutions for the improvement of this flow are essential. This can only be accomplished if we manage to transform passive infrastructure assets into cyber-physical systems. Furthermore, digital twins bring the opportunity to turn passive infrastructure assets into data-centric systems of systems [30]. A digital twin as presented in this work can be the foundation for this advancement.

Building new services based on these data-centric integrations is another advantage of working with a digital twin and integrated (multiple) digital twins. Through application components and networked communication protocols, service-oriented architectures have the capacity to offer services to other systems. Using these services, passive assets are transformed into interactive, living digital twins that gather data, execute analytical models, direct actions, and feed information about all these processes to enable improved decisions for individual systems as well as the system of systems. Conversely, the implementation of a service-oriented digital twin as presented in this book creates opportunities for using a variety of platforms to boost operational effectiveness and creates new tracks for interdisciplinary collaboration and information sharing to support in-depth analyses of smart infrastructure systems [30].

Service interfaces, service implementations, a service bus, and service consumers make up the service-oriented architecture. The data associated with the services can either be shared with specific stakeholders on-demand to enable them to derive and deliver services, or the services can be implemented as an optional extension of a digital twin where the data analytics, software components, and applications run at the higher digital twin level and the results are shared with the consumers of services. In any instance, the parties must agree on the appropriate security, privacy, safety, and similar considerations because the digital twin is understood as a service broker [30].

Even though the digital twin's current implementation does not include integration with other digital twins or provide any service, the service-oriented architecture is a beneficial way of designing the system. One example service that it could provide is an assignment of operational data to the singular parts and units and their usage for process analytics and optimization regularly by additional software components. Asset managers would find this information helpful in making decisions. The next crucial step in making better, smarter infrastructure decisions is unquestionably to offer services to manage energy consumption, environmental pollution, safety, traffic, grid, or sustainability decisions [30].

Additionally, the data management of the digital twin implementation is a vital necessity (see previous section) for a successful implementation. Information from various data sources is intended to be saved in raw, analyzed, and ready-to-share formats and will be stored in both rational and cloud-based databases, local and remote repositories. Stakeholders should discuss data availability, accessibility, challenges related to heterogeneity, and quality in order to prepare an extensive data management plan for the asset. Furthermore, the reference digital twin has been found to be very beneficial to help structuring different data stores and their roles and helped to identify integration-related requirements and needs, in particular, if digital twin impacts operations with exact, predefined work cycles (e.g., food processing [31]). The reuse and compilation of raw and processed data highlight the importance of acquiring a systems perspective when designing digital twins today to enable interoperable systems of systems in the future.

10.3 Synchronization with Piping and Instrumentation Diagram

Due to the statutory provisions, the piping and instrumentation diagram (P&ID) remains a mandatory constituent of the plant documentation and an important data source for the digital twin. The importance of consistency between P&ID and digital twin was already highlighted in Sects. 2.3.1, 6.4.1.2, 7.5.2, and 9.6 which supposes the availability of P&ID in the same CAD system (e.g., AVEVA E3D). Nevertheless, a significant amount of P&ID is still available as scans and, therefore, not suitable for further automated processing by CAD software. To fully support the digital twin and reduce the manual efforts, a transition from existing pixel-oriented diagrams to a machine-readable representation including the spatial and functional interrelationship of all components as well as their structural relationship is necessary. This helps to create an Industry 4.0 conform P&ID diagram [16].

The problem is two-fold: (1) the development of a procedure for the (partially) automated recognition of P&ID from scans of paper documents [32] and (2) the extraction of information about their intercorrelation for real-world applications [33]. Automated interpretation of paper drawings was already applied in previous decades, while results heavily rely on the underlying drawing quality.

More recently, approaches using artificial intelligence are also used here. This approach extracts the required information from P&IDs, creates a machine-readable model of the system automatically, and lists semi-automatically the specific process components and pipeline connections wanted to be included in the digital model of the system. With this list, an intermediate model-based representation can be built, comprising information about different process components, pipeline connections, and attributes related to components like position, rotation, type, and source file [34].

This approach consists of three basic steps: Find symbols of components, find connecting lines (including intersections), and combine the extracted information into a machine-readable representation. Although it is a promising approach, it still lacks important capabilities and robustness to apply to real-world use cases. Due to privacy issues, only very few real diagrams are available for supervised training of AI applications which induces the risk of overfitting [35]. Especially the number of symbols, but also the capability and robustness of line and intersection detection are the main limiting factors, followed by commonly seen problems with noise such as manually added text, folding lines, or dirt on scanned images are not taken into consideration [32].

While no uniform naming conventions throughout the lifecycle of industrial process plants are used, an appropriate technique is necessary for straightforward identification of the same process component from the various (e.g., 2D and 3D) information sources. In our current solution, this task is made optionally by the cooperation partner (as described in Sect. 6.4.1.2). An automated solution could identify corresponding process components from the 3D and 2D models. This identification can be conducted by a structured comparison of 2D with 3D data which can be effectively applied to identify the corresponding components [34]. P&IDs can be analyzed regarding structure and connectivity with various methods [36]. Model comparison can be reliably used to match major process components such as tanks and pumps in the 2D and 3D models, while the accurate match of smaller process components such as pipes must be developed.

A metric to measure the degree of confidence with which particular portions of the 2D and 3D models have been matched must also be developed keeping in mind that the state "as-planned" and "as-built" may differ to a significant extent [37]. A plant's P&ID describes the "as-planned" status or, if properly adjusted during commissioning and handover, the "as-built" status, whereas the 3D model resembles the "as-is" status, including a variety of changes that occur during operation and maintenance and that are rarely properly documented in 2D models [16].

10.4 Discussion

The objective of this book was to develop a new commercial service based on a digital twin of a process plant generated on a scan. In the precedent Chapters, we have generically described this development, from requirements to the implemented solution and the commercial, modular offering that has been accepted by the first customers. Nevertheless, the overall development continues, and a list of challenges still exists (Table 10.2). These challenges are both related to the gaps determined during the development of our solution and to the recent scientific publications.

It should be brought back that requirements management is a dynamic and creative activity when introducing and organizing a rapidly emerging technology such as the digital twin. The user company will hardly have the capacity to respond quickly enough to deal with design and requirement changes if the process is too formalized. It stands to reason that it must be adaptable to diverse project types. It is still unknown to the organization what amount of process formalization is appropriate [38]. This gives the traceability of requirements particular importance (challenge 10 in Table 10.2).

If the company is actively moving more into a model-based definition, it seems natural that the digital models, such as the CAD, FEA, and CAE are managed by a comprehensive PLM or ALM (Asset Lifecycle Management), which is expressed in challenge 9. Supposing appropriate process constraints, it is certainly technically

Table 10.2 Challenges for further development of digital twin, derived from [16]

Nr	Challenge	Reference
1	What are the limitations of component identification methods from point clouds in a cluttered plant environment with complex geometries such as long pipes with many changes of direction?	[40]
2	To what extent and accuracy can legacy P&ID digitalization be automated with respect to small graphical elements, lines, and text?	[36]
3	What is the most suitable abstraction level for comparison of 3D and 2D design information?	[36]
4	What kind of metric could be defined for the confidence of digitalization of 3D and 2D brownfield plant information?	[41]
5	How can dynamic process simulation models be generated in a standards-compliant way?	[34]
6	What architecture can be used to non-disruptively integrate a digital twin to brownfield plants with various degrees of Industry 4.0 readiness in their legacy systems?	[26, 27]
7	How should intellectual property protection be ensured for the development and operation of a digital twin?	[42]
8	In what process operating regions are soft sensors from a digital twin trustworthy?	[43]
9	How can digital twins be included in a Digital Asset Management concept?	[44]
10	How can formalized requirements traceability be generated as a byproduct of Steps 1–9?	[45]

possible to link those to requirements and validation documents in the PLM system. Challenges 6 and 8 related to architecture should be tackled considering the procedure described in Sect. 4.2.4 (Figs. 4.4 and 4.5). The systems engineering approach means that the requirements are broken down and assigned to the components of the plant in a very dynamic way during the development of the digital twin [39].

Considering that the process runs global among several partners, the protection of intellectual property (challenge 7) must be ensured in an appropriate way [41]. The generation of dynamic process simulation (challenge 5) models can be supported by working in ALM, questions on how and to what granularity it is reasonable to represent them in the ALM systems need to be resolved.

Challenges 1, 2, and 3 belong to the functional improvements and, therefore, can be tackled by improvements of algorithms and procedures which is a continuous process that is foreseen in our architecture and workflow (Sects. 6.2.1 and 7.5). There is also a need to feed the operational and quality data from current production upstream into new development projects. The data need to be analyzed so that process and software development can understand what and where the service needs to be improved.

Finally, this entire development presented in this book aims for the replacement of manual efforts. An appropriate metric should be defined and implemented for the confidence of digitalization of 3D and 2D brownfield plant information (challenge 4) which preferably should be implemented as an automated function [40].

10.5 Closing Remarks and Conclusions

An innovative method for the generation of a digital twin of an existing process plant or a part of it in a generic way, including optional elements that are expected to become core to the value proposition in the near future, was the goal of this book. Although it is based on a reimplementation and functional extension of previously developed methods and tools, this book completes this topic independently with a step-by-step description for productive use. Furthermore, the proposed approach helps customers from the industry to acquire their individual digital twins as a service from a well-introduced platform ecosystem [46].

The book presents its content in a scholarly structure. It aimed at a scientific introduction to digital twins in the process industry with a practical outcome and at discussing the emerging role of information technologies in the process industry. Each chapter is meant a unit in itself and dedicated to a specific topic. Readers are encouraged to conceive of these information artifacts—requirements, charts, models, tools, and recommendations—as an initial aid to understanding interrelations and as a foundation for discussion and original contributions regarding specific problems.

As it has become evident, the digital twin is a comprehensive approach to digitization for tackling complex, real-world problems. A specific application for a dedicated purpose on a customer's premise is described in this book. While our digital twins' designs and inputs change continuously, the twins must adapt to the new conditions.

The solution for this challenge is comprised in our approach. For further investigation, the reader will find hundreds of references in the bibliography categorized per chapter, meaning that material selected primarily from first-class scientific journals and scientific books published in the past few years related to the specific scope of a dedicated chapter.

While the digital twin has its lifecycle, an integration into ALM is necessary [44]. Although this was not in the scope of this book, a lifecycle view must be considered for a digital twin in a further step for several reasons [47]. Short working cycles require maturity model development such as comparison with existing models which eliminates the iterative nature of development and contributes to digital maturity [48]. True integration is still a challenging objective, and until the industry reaches a level of maturity and comprehension, additional research and development will be needed to ensure the successful and secure integration of digital twins.

Usually, the product lifecycle is distributed across the globe: the construction and maintenance of a plant involve parties from different locations. Users, their organization, methods, and tools face significant challenges in terms of communication, efficiency, and interoperability as a result of this nearly limitless dispersion [49]. Interoperability is now an indispensable prerequisite due to the synchronized availability of physical and digital products.

Furthermore, aspects related to human–machine interaction were not considered in the focus of this book. Both people using digital twins and those interacting with them place a high priority on security and privacy. The privacy and security risks associated with digital twin deployment are manifold. Overcoming technical, legal, and cultural obstacles is a prerequisite for any significant shift in an industry. By disrupting technologies like the digital twin, this is especially true. Cultures must shift, new methods must be learned, and new skills must be acquired [50].

Finally, seamless digitalization and automation demand ongoing technological leadership and a clearly defined vision expressed in the form of a digital strategy. The concept of the digital twin is more limited by our biased imagination than by available technology. An initial step towards a comprehensive digital twin of a process plant could be the presented solution. In the context of modern IT technology, the next steps could lead towards sensing, learning, and reasoning e.g., by using methods and techniques of artificial intelligence.

References

1. Stjepandić J, Sommer M, Denkena B (2022) DigiTwin: an approach for production process optimization in a built environment. Springer International Publishing Switzerland. https://doi.org/10.1007/978-3-030-77539-1
2. Rupčić N, Majić T, Stjepandić J (2020) Emergence of business ecosystems by transformation of platforms through the process of organizational learning. J Ind Intg Manag 5(2):181–203. https://doi.org/10.1142/S2424862220500086

3. Jones D, Snider C, Nassehi A, Yon J, Hicks B (2020) Characterising the digital twin: a systematic literature review. CIRP J Manuf Sci Technol 29:36–52. https://doi.org/10.1016/j.cirpj.2020.02.002
4. van der Valk H, Haße H, Möller F, Arbter M, Henning JL, Otto B (2020) A taxonomy of digital twins. In: Anderson B, Thatcher J, Meservy R (eds) Proceedings 26th Americas conference on information systems, pp 1–10
5. Jiang Z, Guo Y, Wang Z (2021) Digital twin to improve the virtual-real integration of industrial IoT. J Ind Inf Integr 22:100196. https://doi.org/10.1016/j.jii.2020.100196
6. Li L, Lei B, Mao C (2022) Digital twin in smart manufacturing. J Ind Inf Integr 26:100289. https://doi.org/10.1016/j.jii.2021.100289
7. Davila Delgado JM, Oyedele L (2021) Digital twins for the built environment: learning from conceptual and process models in manufacturing. Adv Eng Inform 49:101332. https://doi.org/10.1016/j.aei.2021.101332
8. Kremer P, Lützenberger J, Müller F, Stjepandić J (2022) An approach for the incremental update of a digital twin of a process plant. Advances in Transdisciplinary Engineering 28:310–319. https://doi.org/10.3233/ATDE220660
9. Kuhn O, Liese H, Stjepandić (2011) Methodology for knowledge-based engineering template update. In: Cavallucci D, Guio R, Cascini G (eds) Building innovation pipelines through computer-aided innovation. Springer, Berlin, pp 178–191. https://doi.org/10.1007/978-3-642-22182-8_14
10. Wahid A, Zhu J, Mauceri S, Li L, Liu M (2023). Digital twins: modelling languages comparison. In: Nicosia G et al (eds) Machine learning, optimization, and data science. LOD 2022. Lecture notes in computer science, vol 13811. Springer, Cham, pp 169–178. https://doi.org/10.1007/978-3-031-25891-6_13
11. Sommer M, Seiffert K (2021) Scan methods and tools for reconstruction of built environments as basis for digital twins. In: Stjepandić J et al (eds) DigiTwin: an approach for production process optimization in a built environment. Springer Nature, pp 51–78. https://doi.org/10.1007/978-3-030-77539-1_4
12. Erkoyuncu JA, del Amo IF, Ariansyah D, Bulk D, Vrabic R, Roy R (2020) A design framework for adaptive digital twins. CIRP Ann 69(1):145–148
13. Leng J, Wang D, Shen W, Li X, Liu Q, Chen X (2021) Digital twins-based smart manufacturing system design in industry 4.0: a review. J Manuf Syst 60:119–137. https://doi.org/10.1016/j.jmsy.2021.05.011
14. Confalone GC, Smits J, Kinnare T (2023) 3D scanning for advanced manufacturing, design, and construction. Wiley, Hoboken. https://doi.org/10.1002/9781119758532
15. Sharma A, Kosasih E, Zhang J, Brintrup A, Calinescu A (2022) Digital twins: state of the art theory and practice, challenges, and open research questions. J Ind Inf Integr 30:100383. https://doi.org/10.1016/j.jii.2022.100383
16. Sierla S, Azangoo M, Rainio K, Papakonstantinou N, Fay A, Honkamaa P, Vyatkin V (2022) Roadmap to semi-automatic generation of digital twins for brownfield process plants. J Ind Inf Integr 27:100282. https://doi.org/10.1016/j.jii.2021.100282
17. Segovia M, Garcia-Alfaro J (2022) Design, modeling and implementation of digital twins. Sensors 22:5396. https://doi.org/10.3390/s22145396
18. Semeraro C, Lezoche M, Panetto H, Dassisti M (2023) Data-driven invariant modelling patterns for digital twin design. J Ind Inf Integr 31:100424. https://doi.org/10.1016/j.jii.2022.100424
19. Anbari E, Parvini M (2022) Process plant layout optimization considering domino effects and safety devices in a large-scale chemical plant. Comput Chem Eng 167:108006. https://doi.org/10.1016/j.compchemeng.2022.108006
20. Guo H, Zhu Y, Zhang Y, Ren Y, Chen M, Zhang R (2021) A digital twin-based layout optimization method for discrete manufacturing workshop. Int J Adv Manuf Technol 112:1307–1318. https://doi.org/10.1007/s00170-020-06568-0
21. Xu LD (2020) Industrial information integration—an emerging subject in industrialization and informatization process. J Ind Inf Integr 17:100128. https://doi.org/10.1016/j.jii.2020.100128

22. Liu S, Lu Y, Shen X, Bao J (2023) A digital thread-driven distributed collaboration mechanism between digital twin manufacturing units. J Manuf Syst 68:145–159. https://doi.org/10.1016/j.jmsy.2023.02.014
23. Riascos R, Majić T, Ostrosi E, Sagot JC, Stjepandić J (2022) Integrated multilayer architecture with multi interface entity model for risk management in modular product design. Procedia CIRP 109:647–652. https://doi.org/10.1016/j.procir.2022.05.308
24. López A, Ogayar CJ, Jurado JM, Feito FR (2023) Metaheuristics for the optimization of terrestrial LiDAR set-up. Autom Constr 146:104675. https://doi.org/10.1016/j.autcon.2022.104675
25. DTDL models—Azure Digital Twins. Microsoft Docs (2022) https://docs.microsoft.com/en-us/azure/digital-twins/concepts-models. Accessed 15 April 2023
26. Zhang L, Zhou L, Horn BKP (2021) Building a right digital twin with model engineering. J Manuf Syst 59:151–164. https://doi.org/10.1016/j.jmsy.2021.02.009
27. Tao F, Xiao B, Qi Q, Cheng J, Ji P (2022) Digital twin modeling. J Manuf Syst 64:372–389. https://doi.org/10.1016/j.jmsy.2022.06.015
28. Son YH, Kim G-Y, Kim HC, Jun C, Noh SD (2022) Past, present, and future research of digital twin for smart manufacturing. Journal of Computational Design and Engineering 9(1):1–23. https://doi.org/10.1093/jcde/qwab067
29. Jiang L, Su S, Pei X, Chu C, Yuan Y, Wang K (2023) Product-part level digital twin modeling method for digital thread framework. Comput Ind Eng 179:109168. https://doi.org/10.1016/j.cie.2023.109168
30. Gürdür Broo D, Bravo-Haro M, Schooling J (2022) Design and implementation of a smart infrastructure digital twin. Autom Constr 136:104171. https://doi.org/10.1016/j.autcon.2022.104171
31. Liu X, Jiang D, Tao B, Xiang F, Jiang G, Sun Y, Kong J, Li G (2023) A systematic review of digital twin about physical entities, virtual models, twin data, and applications. Adv Eng Inform 55:101876. https://doi.org/10.1016/j.aei.2023.101876
32. Mertens N, Wohlfahrt T, Hartmann N, Reddy CBV (2023) Automatic transformation of HVAC diagrams into machine-readable format. In: Noël F, Nyffenegger F, Rivest L, Bouras A (eds) Product lifecycle management. PLM in transition times: the place of humans and transformative technologies. PLM 2022. IFIP advances in information and communication technology, vol 667. Springer, Cham, pp 410–419. https://doi.org/10.1007/978-3-031-25182-5_40
33. Bamberg A, Urbas L, Bröcker S, Bortz M, Kockmann N (2021) The digital twin—your ingenious companion for process engineering and smart production. Chem Eng Technol 44:954–961. https://doi.org/10.1002/ceat.202000562
34. Martínez GS, Sierla SA, Karhela TA, Lappalainen J, Vyatkin V (2018) Automatic generation of a high-fidelity dynamic thermal-hydraulic process simulation model from a 3D plant model. IEEE Access 6:45217–45232. https://doi.org/10.1109/ACCESS.2018.2865206
35. Sommer M, Stjepandić J, Stobrawa S, von Soden M (2023) Automated generation of digital twin for a built environment using scan and object detection as input for production planning. J Ind Inf Integr 33:100462. https://doi.org/10.1016/j.jii.2023.100462
36. Rantala M, Niemistö H, Karhela T, Sierla S, Vyatkin V (2019) Applying graph matching techniques to enhance reuse of plant design information. Comput Ind 107:81–98. https://doi.org/10.1016/j.compind.2019.01.005
37. Grau M, Korol W, Lützenberger J, Stjepandić J (2021) Automated generation of a digital twin of a process plant by using 3D scan and artificial intelligence. Advances in Transdisciplinary Engineering 16:93–102. https://doi.org/10.3233/ATDE210087
38. Stolt R, Andre S (2022) Requirements handling in multidisciplinary product development—a company study. Advances in Transdisciplinary Engineering 28:115–124. https://doi.org/10.3233/ATDE220638
39. Verhagen WJC, Stjepandić J, Wognum N (2019) Future perspectives in systems engineering. In: Stjepandić J, Wognum N, Verhagen WJC (eds) Systems engineering in research and industrial practice. Springer International Publishing, pp 403–420. https://doi.org/10.1007/978-3-030-33312-6_14

40. Xie Y, Li S, Liu T, Cai Y (2023) As-built BIM reconstruction of piping systems using PipeNet. Autom Constr 147:104735. https://doi.org/10.1016/j.autcon.2022.104735
41. Wang Q, Li J, Tang X, Zhang X (2022) How data quality affects model quality in scan-to-BIM: a case study of MEP scenes. Autom Constr 144:104598. https://doi.org/10.1016/j.autcon.2022.104598
42. Biahmou A, Stjepandić J (2016) Towards agile enterprise rights management in engineering collaboration. International Journal of Agile Systems and Management 9:302–325. https://doi.org/10.1504/IJASM.2016.081564
43. Li D, Liu J, Zeng Y, Cheng G, Dong B, Chen YF (2022) 3D model-based scan planning for space frame structures considering site conditions. Autom Constr 140:104363. https://doi.org/10.1016/j.autcon.2022.104363
44. NN (2022) AVEVA™ asset information management discovery. https://www.aveva.com/en/solutions/operations/asset-visualization/. Accessed 22 Nov 2022
45. Mirzaei K, Arashpour M, Asadi E, Masoumi H, Bai Y, Behnood A (2022) 3D point cloud data processing with machine learning for construction and infrastructure applications: a comprehensive review. Adv Eng Inform 51:101501. https://doi.org/10.1016/j.aei.2021.101501
46. Stjepandić J, Sommer M, Stobrawa S (2022) The commercialization of digital twin by an extension of a business ecosystem. In Stjepandić J et al (eds) DigiTwin: an approach for production process optimization in a built environment. Springer Nature, pp 205–233. https://doi.org/10.1007/978-3-030-77539-1_10
47. Hohmann L, Kössl K, Kockmann N, Schembecker G, Bramsiepe C (2017) Modules in process industry—a life cycle definition. Chem Eng Process 111:115–126. https://doi.org/10.1016/j.cep.2016.09.017
48. Perera S, Jin X, Das P, Gunasekara K, Samaratunga M (2023) A strategic framework for digital maturity of design and construction through a systematic review and application. J Ind Inf Integr 31:100413. https://doi.org/10.1016/j.jii.2022.100413
49. Paniagua C, Delsing J (2022) Autonomous runtime consumer interface generation and deployment for service interoperability. J Ind Inf Integr 28:100355. https://doi.org/10.1016/j.jii.2022.100355
50. Agrawal A, Thiel R, Jain P, Singh V, Fischer M (2023) Digital twin: where do humans fit in? Autom Constr 148:104749. https://doi.org/10.1016/j.autcon.2023.104749

Correction to: Generation and Update of a Digital Twin in a Process

Correction to:
J. Stjepandić et al., ***Generation and Update of a Digital Twin in a Process Plant*****, https://doi.org/10.1007/978-3-031-47316-6**

In the original version of the book, the following belated corrections were incorporated: Few typographical errors and change in figure positions have been updated in the chapters.The book has been updated with the changes.

The updated version of the book can be found at
https://doi.org/10.1007/978-3-031-47316-6

J. Stjepandić et al., *Generation and Update of a Digital Twin in a Process Plant*,
https://doi.org/10.1007/978-3-031-47316-6_11

MIX
Papier aus verantwortungsvollen Quellen
Paper from responsible sources
FSC® C105338

FSC
www.fsc.org

If you have any concerns about our products,
you can contact us on
ProductSafety@springernature.com

In case Publisher is established outside the EU,
the EU authorized representative is:
Springer Nature Customer Service Center GmbH
Europaplatz 3, 69115 Heidelberg, Germany

Printed by Libri Plureos GmbH
in Hamburg, Germany